MARINE CORPS MANUAL

1921

WASHINGTON
GOVERNMENT PRINTING OFFICE
1922

CHANGES IN MARINE CORPS MANUAL NO. 1.

HEADQUARTERS MARINE CORPS,
Washington, March 14, 1923.

The following changes in the Marine Corps Manual shall be made immediately upon receipt of this order.

JOHN A. LEJEUNE,
Major General Commandant.

Approved:

T. ROOSEVELT,
Acting Secretary of the Navy.

Make the following changes in ink:

Art. 205. Strike out.

Art. 306 (1). After the eleventh line on page 25 insert the following:
2 pajamas, suits.

Art. 604 (3). In the last line change "authorized" to "unauthorized."

Art. 916 (12). In the sixteenth line change the word "known" to "shown."

Art. 1304 (5). In the last line strike out the words "Army appropriation act of June 30, 1921," and insert the words "act of September 22, 1922."

Art. 1318 (1). In the seventh line, before the word "certificate," insert the word "discharge."

Art. 1318 (3). In the first line strike out the words "in a similar manner" and insert in lieu thereof the words "by the man's immediate commanding officer."

Art. 1322 (5). In the fourth line, after the word "of" insert the words "acceptance for," and after the word "enlistment," same line, strike out the words "or bona fide home or residence, whichever is elected."

Art. 1329 (3). Strike out.

Index page 148. Insert the following entry:

Enlisted men, arrival in U. S. from foreign station 934

Index page 149. Insert:

Enlisted men, time lost, entry of, in service-record book 935, 936

Index page 154. Insert:

Guard duty and drills at navy yards and naval stations 533

Index page 155. Insert:

Insignia, when withheld 1013

Index page 157. Insert:

Mess, detachment, cashbook for 1124

Index page 165. Insert:

Staff sergeants 405

Index page 167. Insert:

Time lost, entry of, in service-record book 935, 936

40677—23

The following changes are made in the accompanying pages of the Marine Corps Manual. These pages will be inserted in their proper places.

Art. 8. Modified and reprinted on page 2.
Art. 206 (1). Modified and reprinted on page 21.
Art. 206 (9). New paragraph inserted and printed on page 22.
Art. 207. Modified and reprinted on page 22.
Art. 404. Modified and reprinted on pages 27 and 28.
Art. 405. Modified and reprinted on page 28.
Art. 508 (2). Modified and reprinted on page 36.
Arts. 519, 520, and 522. Modified and reprinted on pages 38 and 38a.
Art. 530 (3), (8), (21), (22), (23), (27), (28), (32), and (40). Modified and reprinted on pages 40–44.
Art. 532 (2). Modified and reprinted on page 45.
Art. 533. New article inserted and printed on page 45.
Art. 607 (5). Modified and reprinted on page 55.
Art. 608 (9). Modified and reprinted on page 56.
Arts. 615 (3) and 616 (1). Modified and reprinted on page 58.
Arts. 934, 935, and 936. New articles inserted and printed on page 84.
Art. 1004 (1). Modified and reprinted on page 85.
Art. 1013. New article inserted and printed on page 89.
Art. 1124. New article inserted and printed on page 94.
Art. 1304. Modified and reprinted on page 126.

TABLE OF CONTENTS.

Changes in Marine Corps Manual.

[Enter number and dates of changes as indicated.]

C. M. C. M. No.	Dated.	Effective.	Date made.

MARINE CORPS MANUAL.

HEADQUARTERS U. S. MARINE CORPS,
Washington, December 31, 1921.

The accompanying Marine Corps Manual is for the instruction and guidance of the United States Marine Corps and supersedes Marine Corps Orders of prior date, except No. 46, series of 1921.

JOHN A. LEJEUNE,
Major General Commandant.

Approved:

EDWIN DENBY,
Secretary of the Navy.

CHAPTER 1.

GENERAL INSTRUCTIONS.

CORRESPONDENCE.

1

Addressed to Major General Commandant.—All communications relating to details of officers, transfers, applications for orders or for revocation thereof, as well as applications for leaves of absence and furloughs which are not granted by other authority, and applications for extensions of such leaves and furloughs beyond 30 days shall be addressed to the Major General Commandant and forwarded through official channels.

2

From officers at Headquarters.—All official communications between officers of the Corps stationed at Headquarters and other officers (except letters containing money and checks, and such routine correspondence by or with officers of or between the staff departments as requires no action of the Major General Commandant) shall be forwarded through that officer.

3

General officers of the Marine Corps commanding departments, posts, stations, brigades, or other organizations of the Marine Corps, will be addressed, and will style themselves in all official correspondence as "The Commanding General." followed by the name of the department, post, station, etc.

4

Names in full.—To avoid error, confusion, and delay in conducting correspondence, official communications, other than those by wire, containing the names of officers or enlisted men shall in every instance give both the Christian and the surname of such officers and enlisted men written in full, except middle names, for which initials only may be used. In case there is more than one enlisted man of the same name at a given post or station, the one to whom it is intended to refer must be identified by giving the place and date of his enlistment, or otherwise.

5

(1) *Registered mail.*—The Post Office Department considers that its responsibility ceases with the delivery of registered mail to a duly authorized mail orderly, and that in the event of its loss after delivery to him the sender must look to the naval authorities for reimbursement.

(2) Commanding officers will require every person handling registered mail to receipt for it, making sure that such mail is delivered to and signed for by the person to whom it is addressed. Registry receipt books will be kept for this purpose.

6

Personal mail.—When mail is received at a post for an officer or enlisted man who has just been transferred, it will be redirected and forwarded to the new station of the officer or enlisted man concerned and not returned to the sender or forwarded to the Adjutant and Inspector.

OFFICIAL ORDERS, ETC.

7

File of.—The commanding officer of marines shall keep in his office a file of all orders and circulars relating to his command and shall cause all general orders affecting the officers and enlisted men to be published at the first parade following the receipt of such orders.

8

(1) *Distribution.*—Commanding officers of Marine Corps posts and organizations will see that all officers under their command are furnished with copies of orders issued from Headquarters.

(2) If the supply of orders issued is not sufficient, report should be made at once in order that the mailing list may be corrected.

(3) Orders shall be distributed by commanding officers as follows:

(*a*) Marine Corps Orders, circular letters, and changes in Marine Corps publications: To all officers, commanding officers' offices, staff offices, and companies.

(*b*) Navy Department General Orders, changes in Navy Department publications, court-martial orders, changes in Army Regulations: To all officers, commanding officers' offices, and staff offices.

(*c*) War Department General Orders: To commanding officers' offices and staff offices.

(*d*) Marksmanship qualifications: To commanding officers' offices and assistant paymasters.

DUTIES OF OFFICERS.

9

Officers serving with a command shall familiarize themselves with the details of its administration, and shall perform such duties connected therewith as may be prescribed by the commanding officer.

REPORT OF ADDRESS.

10

On leave.—An officer of the Marine Corps on leave of absence granted by the Major General Commandant, or on the retired list, shall report his address and change of residence to the Major General Commandant.

11

While delaying on proceed orders.—An officer who takes advantage of any delay allowed him on "proceed" or "proceed without delay" travel orders will leave at his last station his telegraphic and mail addresses during the delay period.

AVIATION OFFICER PERSONNEL.

12

(1) Officers detailed to aviation duty will be detailed for a period of five years unless sooner relieved.

(2) Officers showing marked aptitude for aviation duty may be redetailed after a period of three years or more with troops in order that officers of command experience and field rank may be available for aviation.

(3) In submitting fitness reports of officers on aviation duty, commanding officers will note thereon the desirability of a redetail to aviation duty of the officer concerned, giving particular consideration to the officer's ability for either executive or command assignments in aviation.

Relations Between Officers and Men.

15

(1) *Comradeship and brotherhood.*—The World War wrought a great change in the relations between officers and enlisted men in the military services. A spirit of comradeship and brotherhood in arms came into being in the training camps and on the battlefields. This spirit is too fine a thing to be allowed to die. It must be fostered and kept alive and made the moving force in all Marine Corps organizations.

(2) *Teacher and scholar.*—The relation between officers and enlisted men should in no sense be that of superior and inferior nor that of master and servant, but rather that of teacher and scholar. In fact, it should partake of the nature of the relation between father and son, to the extent that officers, especially commanding officers, are responsible for the physical, mental, and moral welfare, as well as the discipline and military training of the young men under their command who are serving the nation in the Marine Corps.

(3) *The realization* of this responsibility on the part of officers is vital to the well-being of the Marine Corps. It is especially so, for the reason that so large a proportion of the men enlisting are under twenty-one years of age. These men are in the formative period of their lives, and officers owe it to them, to their parents, and to the nation, that when discharged from the service they should be far better men physically, mentally, and morally than they were when they enlisted.

(4) *To accomplish this task* successfully a constant effort must be made by all officers to fill each day with useful and interesting instruction and wholesome entertainment for the men. This effort must be intelligent and not perfunctory, the object being not only to do away with idleness, but to train and cultivate the bodies, the minds, and the spirit of our men.

(5) *Love of corps and country.*—To be more specific, it will be necessary for officers not only to devote their close attention to the many questions affecting the comfort, health, military training and discipline of the men under their command, but also actively to promote athletics and to endeavor to enlist the interest of their men in building up and maintaining their bodies in the finest physical condition; to encourage them to enroll in the Marine Corps Institute and to keep up their studies after enrollment; and to make every effort by means of historical, educational and patriotic addresses to cultivate in their hearts a deep abiding love of the corps and country.

(6) *Leadership.*—Finally, it must be kept in mind that the American soldier responds quickly and readily to the exhibition of qualities of leadership on the part of his officers. Some of these qualities are industry, energy, initiative, determination, enthusiasm, firmness, kindness, justness, self-control, unselfishness, honor, and courage. Every officer should endeavor by all means in his power to make himself the possessor of these qualities and thereby to fit himself to be a real leader of men.

Enlisted Men.

16

Whenever an enlisted man is injured or becomes ill special effort will be made to ascertain the name and address of his next of kin.

17

Enlisted men should be advised of their privilege to request correction of any erroneous entry appearing in their enlistment papers or service-record books.

18

The detail of an enlisted man in the dual capacity of commissary sergeant and mess steward is prohibited.

19

(1) *Employment as servants.*—Article 553, U. S. Navy Regulations, provides that: "Under no circumstances shall any enlisted man be employed as a servant."

(2) At all marine barracks and Marine Corps posts and offices within and without the continental limits of the United States this regulation will be interpreted to have the following meaning, viz:

(*a*) No enlisted man will be employed by officers to perform any duty which in civil life is performed by a man or woman employed as a servant.

(*b*) Enlisted men will not be employed as chauffeurs of privately owned automobiles.

(*c*) Enlisted men may be used to deliver horses to officers' quarters and to call for same.

(*d*) Coal, and wood of desired lengths, commissary and other supplies will be delivered to quarters of officers and enlisted men.

(*e*) Firemen will be detailed to care for the heating plants of public quarters, not more than one man to be employed for each five sets of quarters. These men will perform their duties under the post quartermaster and will not be requested or ordered to perform any work in or around the quarters by officers occupying the quarters or any members of their families.

(*f*) Officers on commutation of quarters status will not be allowed the services of an enlisted man in their quarters in any capacity.

(*g*) The necessary repairs to, and upkeep of, public quarters will be made by the quartermaster's department.

(*h*) Only those officers commanding administrative units will be entitled to orderlies; such orderlies will be used for official purposes only in connection with the official duties of the officers concerned.

(*i*) Subparagraph 2(*h*) is not intended to apply to general officers occupying public quarters, except in so far as it provides that orderlies assigned to general officers will not be used as servants.

20

Engaging in business.—Hereafter no enlisted man in the active service of the United States in the Army, Navy, and Marine Corps, respectively, whether a noncommissioned officer, musician, or private, shall be detailed, ordered, or permitted to leave his post to engage in any pursuit, business, or performance in civil life for emolument, hire, or otherwise, when the same shall interfere with the customary employment and regular engagement of local civilians in their respective arts, trades, or professions. (Act of June 3, 1916—39 S. 188.)

21

Post and regimental bands, or members thereof, shall not receive remuneration for furnishing music outside the limits of their respective commands when the furnishing of such music places them in competition with local civilian musicians.

22

In acknowledging verbal orders or instructions, enlisted men will use the expression "aye, aye, sir." "Very well, sir," and "yes, sir," or similar expressions will not be used for this purpose.

23

Hair cuts.—Enlisted men will, at all times, wear the hair neatly and closely trimmed. The hair may be clipped at the edges of the sides and back, but must be so trimmed as to present an evenly graduated appearance, and must not be over one inch in length on top of the head and above the forehead. The back of the neck must not be shaved.

Beneficiary Slips.

24

(1) *Dependent relatives.*—Commanding and other officers will impress on enlisted men the necessity for designation of dependent relatives. The death gratuity can not be paid unless the names of dependent relatives have been duly recorded.

(2) *A new beneficiary slip* is not required upon reenlistment unless there is a change of beneficiary.

Identification Tags.

25

(1) Two identification tags will be issued to each officer and enlisted man of the Marine Corps, to be worn when engaged in field service. One tag will be suspended from the neck underneath the clothing by a cord or thong passed through the small hole in the tag, the second tag to be suspended from the first one by a short piece of string or tape. These tags are prescribed as a part of the uniform, and when not worn as directed herein will be habitually kept in the possession of the owner. When not worn they will be regarded as part of the field kit and will be regularly inspected.

(2) In order to secure the proper interment of those who fall in battle, and to establish beyond a doubt their identity, should it become desirable subsequently to disinter the remains for removal to a national or post cemetery or for shipment home, the identification tag suspended from the neck of the officer or enlisted man will in all cases be interred with the body. The duplicate tag attached thereto will be removed at the time of burial and turned over to the surgeon or person in charge of the burial, from which a record of same, together with the cause and date of death, shall be made and reported to the commanding officer.

(3) These tags will be stamped as follows: Officers, full name and rank at date of issue; enlisted men, full name and date of first enlistment in the Marine Corps, the tags of both officers and enlisted men to have the letters "U. S. M. C." plainly stamped thereon.

(4) Tags for men enlisting will be stamped and issued at the recruit depots, except in the case of reenlisted men who have not previously been supplied with them. In such cases tags will be stamped and issued at the posts to which the men are transferred.

(5) The original issue of tags and tape will be made gratuitously, but issues made to replace those lost will be checked against the man's pay account.

(6) The Secretary of the Navy has authorized the use of the Marine Corps identification tag until the exhaustion of the present supply, after which the tag prescribed in the Navy Regulations will be used.

MESS STEWARDS, COOKS, MESSMEN.

27

Mess stewards.—Where organizations or detachments are merged into a general mess or where there is no company organization, additional compensation may be given a mess steward, one for each general mess. Mess stewards shall be detailed from the grade of sergeant or below.

28

(1) *Details.*—At regularly established posts and other shore stations all details as mess stewards, cooks, and messmen shall be made on the first day of the month; and the number detailed shall be based on the ration strength of the command on that date. No further detail of cooks or messmen will be permitted on intermediate days of the month, except to fill a vacancy caused by sickness, confinement, promotion, transfer, discharge, death, or desertion, or where the ration strength of the command is increased by 25 per cent of the number as shown on the 1st day of the month, nor thereafter except upon a like increase. For all commands whose ration strength on the 1st day of the month is 400 or more there will be allowed additional details of cooks and messmen for each 100 increase in ration strength during the month.

(2) *Ships.*—Commanding officers of ships are authorized to assign privates of marines to duty as cooks.

29

(1) *Cooks, allowance.*—Cooks shall invariably be detailed from the grades of private and private first class; and in no case shall men receiving extra compensation as such be detailed on or receive additional pay for any other special duty. In detailing cooks the following apportionment shall be observed:

Number of men.	Cooks first class.	Cooks second class.	Cooks third class.	Cooks fourth class.
1. 50 men or under, serving separately	1	....	....	....
2. 51 to 75 men, serving separately	1	....	1	....
3. Less than 75 men, serving as a company	1	....	1	....
4. 76 to 150 men, serving as a company	1	1	....	....
5. 151 to 200 men, serving as a company	1	1	1	....
5a. When over 200 men are serving as a company the garrison allowance will apply	....	....	....	....
6. 100 men or less, serving in garrison	1	1	....	....
7. 101 to 200 men, serving in garrison	1	1	1	....
8. 201 to 300 men, serving in garrison [1]	1	1	1	1
9. 301 to 400 men, serving in garrison [1]	1	1	1	2
10. 401 to 500 men, serving in garrison [1]	1	1	2	2
11. 501 to 600 men, serving in garrison [1]	1	2	2	2
12. 601 to 700 men, serving in garrison [1]	1	2	2	3
13. 701 to 800 men, serving in garrison [1]	1	2	3	3
14. 801 to 900 men, serving in garrison [1]	2	2	3	3
15. 901 to 1,000 men, serving in garrison [1]	2	2	3	4
16. 1,001 to 1,100 men, serving in garrison [1]	2	2	4	4
17. 1,101 to 1,200 men, serving in garrison [1]	2	3	4	4
18. 1,201 to 1,300 men, serving in garrison [1]	2	3	4	5
19. 1,301 to 1,400 men, serving in garrison [1]	2	3	5	5
20. 1,401 to 1,500 men, serving in garrison [1]	2	4	5	5
21. 1,501 to 1,600 men, serving in garrison [1]	2	4	5	6
22. 1,601 to 1,700 men, serving in garrison [1]	2	4	6	6
23. 1,701 to 1,800 men, serving in garrison [1]	2	5	6	6
24. For each additional hundred men over 1,800	....	....	....	1

[1] Except where there are two or more distinct messes, when cooks shall be allowed as indicated above.

(2) *Bakers.*—At places where bread is baked the allowance of additional cooks as bakers shall be as follows:

Number of men.	Cooks first class.	Cooks second class.	Cooks third class.	Cooks fourth class.
1. 100 men or less	1	....	....	....
2. 101 to 300 men	1	....	....	1
3. 301 to 1,000 men	1	1	1	....
4. 1,001 to 1,500 men	1	2	....	1
5. 1,501 to 2,000 men	2	2	....	2
6. 2,001 to 2,500 men	2	2	1	3
7. 2,501 to 3,000 men	2	2	2	3
8. For each 1,000 men additional	1	1	....	1

30

Messmen, not exceeding 1 for every 20 men, shall be detailed from the grades of private and private first class. Under no circumstances shall noncommissioned officers be so detailed. Enlisted men of the Marine Corps detailed as messmen afloat are also entitled to the same extra compensation for service with crew messes as is allowed enlisted men of the Navy under like circumstances.

SIGNALMEN.

31

(1) *Examination by board.*—Candidates for detail as signalmen will be examined by a board of officers (three, if practicable) designated by the appointing power.

(2) *Standards for examination.*—The following standards are adopted for the guidance of boards appointed to examine men for detail as signalmen and will be considered the minimum for qualification in each class:

(*a*) Signalman first class—
  Wigwag: Send and receive 5 words a minute.
  Semaphore: Send and receive 12 words a minute.
  Heliograph, Colt lantern or other flashlight instrument in use in the command: Send and receive 8 words a minute.

(*b*) Signalman second class—
  Wigwag: Send and receive 4 words a minute.
  Semaphore: Send and receive 10 words a minute.
  Heliograph, Colt lantern or other flashlight instrument in use in the command: Send and receive 6 words a minute.

(*c*) Signalman third class—
  Wigwag: Send and receive 3 words a minute.
  Semaphore: Send and receive 8 words a minute.
  Heliograph, Colt lantern or other flashlight instrument in use in the command: Send and receive 4 words a minute.

(*d*) NOTE.—Each five letters in a message will be counted as one word. Candidates must be able to set up and operate heliograph and Colt lantern apparatus. For the test stations will be at least 300 yards apart.

(3) The standard for radio, telegraph, and buzzer operators of the signal companies will include a knowledge of the instruments used and will necessarily be variable. This standard must be decided by the examining board.

32

Complements.—Following are the complements of signalmen authorized for organizations of the Marine Corps:

Organization.	Class of signalman.			Total.
	First.	Second.	Third.	
Marine Barracks, Pearl Harbor, Hawaii	2	2	2	6
Marine Barracks, Naval Station, Cavite, P. I.	2	2	2	6
Marine Barracks, Naval Station, Olongapo, P. I.	1	1	1	3
Marine Detachment, Managua, Nicaragua	1	1	1	3
Marine Barracks, Naval Station, Guantanamo, Cuba	1	1	1	3
Marine Barracks, Naval Station, Guam	4	4	4	12
Marine Barracks, St. Thomas, Virgin Islands	1	1	1	3
Marine Detachment, American Legation, Peking, China	3	2	2	7
First Brigade, Haiti	17	17	17	51
Second Brigade, Santo Domingo	23	23	23	69
Fourth Brigade (while at Quantico):				
Each rifle company	1	1	1	3
A brigade on independent expeditionary duty:				
Brigade headquarters communication platoon	4	4	4	12
Each regimental communication platoon	3	3	3	9
Each battalion communication platoon	2	2	2	6
Each rifle company	1	1	1	3
A regiment on independent expeditionary duty:				
Regimental communication platoon	3	3	3	9
Each battalion communication platoon	2	2	2	6
Each rifle company	1	1	1	3
A battalion on independent expeditionary duty:				
Battalion communication platoon	3	3	3	9
Each rifle company	1	1	1	3
A company on independent expeditionary duty	2	2	2	6

33

(1) *Officers authorized to detail.*—The commanding officer of any command is authorized to fill any vacancies in the authorized complement of signalmen thereof by the detail of men who shall have been found qualified by a board as provided in art. 31.

(2) Details as signalmen will be shown on the muster and pay rolls of the organization next rendered after the details are made.

(3) The detail of a marine as signalman may be terminated at the pleasure of the appointing power, and automatically ceases upon the transfer of the man from the organization in which he was detailed.

SPECIALISTS.

34

(1) *Authority to rate.*—Commanding officers are authorized to rate privates and privates first class as specialists, in numbers and ratings not to exceed the allowance prescribed for their respective commands, except that in computing such allowance for any class of specialists a result including 50 per cent or more of one will be taken to allow the next higher whole number; thus a command having an authorized strength of 50 privates and privates first class and an authorized allowance of third class specialists of 1 per cent of such strength would be allowed one third class specialist. Cooks, bakers, and messmen will not be rated as specialists.

(2) *Posts in United States.*—For a marine barracks or marine post in the continental United States (not including marine detachments), which does not

have a special allowance otherwise authorized, there is authorized an allowance of not to exceed the following percentages of the total authorized strength of privates and privates first class:

	Per cent.
Third class specialists	1
Fourth class specialists	2
Fifth class specialists	1

(3) ***Posts outside United States.***—For a marine barracks or marine post outside the continental limits of the United States (not including marine detachments), which does not have a special allowance otherwise authorized, there is authorized an allowance of not to exceed the following percentages of the total authorized strength of privates and privates first class:

	Per cent.
Third class specialists	1.0
Fourth class specialists	1.5
Fifth class specialists	1.0

(4) ***For the Marine Barracks, Mare Island, Calif.,*** there is authorized an allowance of not to exceed the following percentages of the total authorized strength of privates and privates first class (excluding recruits):

	Per cent.
Third class specialists	1.5
Fourth class specialists	2.0
Fifth class specialists	2.0

(5) ***For the Marine Barracks, Parris Island, S. C.,*** there is authorized an allowance of not to exceed the following percentages of the total authorized strength of privates and privates first class (excluding recruits):

	Per cent.
Third class specialists	1.5
Fourth class specialists	3.0
Fifth class specialists	4.0

(6) ***For the Marine Barracks, Quantico, Va.,*** there is authorized an allowance of not to exceed the following percentages of the total authorized strength of privates and privates first class (excluding the third and fourth brigades):

	Per cent.
Third class specialists	1.5
Fourth class specialists	3.0
Fifth class specialists	4.0

(7) ***For the Marine Barracks, Peking, China,*** there is authorized an allowance of not to exceed the following percentages of the total authorized strength of privates and privates, first class:

	Per cent.
Third class specialists	1
Fourth class specialists	2
Fifth class specialists	1

(8) ***For the First Brigade*** of marines, Haiti, there is authorized an allowance of not to exceed the following percentages of the total authorized strength of privates and privates, first class (in determining the allowance of specialists for the First Brigade, the number of privates and privates, first class, detailed to the Gendarmerie d'Haiti will not be included as part of the authorized strength of the brigade):

	Per cent.
Third class specialists	1
Fourth class specialists	2
Fifth class specialists	2

The brigade commander is authorized to allot allowances to the units of his command as in his opinion the interests of the service demand.

(9) *For the Second Brigade* of marines, Dominican Republic, there is authorized an allowance of not to exceed the following percentages of the total authorized strength of privates and privates, first class (in determining the allowance of specialists for the Second Brigade, the number of privates and privates, first class, detailed to the Policia Nacional Dominicana will not be included as part of the authorized strength of the brigade):

	Per cent.
Third class specialists	1
Fourth class specialists	2
Fifth class specialists	2

The brigade commander is authorized to allot allowances to the units of his command as in his opinion the interests of the service demand.

(10) *For the Tenth Regiment* of marines there is authorized an allowance of not to exceed the following percentages of the total authorized strength of privates and privates, first class:

	Per cent.
Third class specialists	0.7
Fourth class specialists	2.5
Fifth class specialists	2.5

(11) *For the Third Separate Company (Signal)* there is authorized an allowance of not to exceed the following percentages of the total authorized strength of privates and privates, first class:

	Per cent.
Third class specialists	0.7
Fourth class specialists	3.0
Fifth class specialists	3.0

(12) *For the First Separate Battalion (Engineer)* there is authorized an allowance of not to exceed the following percentages of the total authorized strength of privates and privates, first class:

	Per cent.
Third class specialists	0.7
Fourth class specialists	3.0
Fifth class specialists	3.0

(13) *For the Fourth Brigade* of marines, there is authorized an allowance of not to exceed the following percentages of the total authorized strength of privates and privates, first class.

Forward echelon, Brigade headquarters:	Per cent.
Third class specialists	6
Fourth class specialists	4

Each Infantry regiment:	Per cent.
Third class specialists	0.5
Fourth class specialists	1.0
Fifth class specialists	2.0

(14) *For the Marine Detachment, Managua, Nicaragua,* there is authorized an allowance of not to exceed the following percentages of the total authorized strength of privates and privates, first class:

	Per cent.
Third class specialists	1
Fourth class specialists	2
Fifth class specialists	1

(15) *For the Fifth Brigade* of marines there is authorized an allowance of not to exceed the following percentages of the total authorized strength of privates and privates, first class:

Forward echelon, Brigade headquarters:	Per cent.
Third class specialists	6
Fourth class specialists	4

Each Infantry regiment:	Per cent.
Third class specialists	0.5
Fourth class specialists	1.0
Fifth class specialists	2.0

(16) *For Marine Corps aviation*, there is authorized an allowance of not to exceed the following percentages of the total authorized strength of privates and privates, first class; such allowances not to be charged to the allowances authorized for any command to which aviation units may be attached, but to be exclusive of and in addition thereto:

	Per cent.
Third class specialists	7
Fourth class specialists	5
Fifth class specialists	3

The total allowance for Marine Corps aviation shall be allotted to the various aviation units as directed by the officer in charge of Marine Corps aviation, and the commanding officers of such units are authorized to rate specialists within such allotments, subject to instructions from the officer in charge of Marine Corps aviation.

(17) *Bands*.—For commands which have bands authorized by the Major General Commandant the allowance of specialists will be increased by the number and ratings of specialists allowed for such bands. Enlisted men of the grades of private, first class, and private, who are assigned to duty as musicians with Marine Corps bands, except the Marine Band, will be given ratings as musicians, first, second, or third class, with extra monthly pay as specialists, as follows:

Specialists, third class ($15.00).
  Musicians, first class.
Specialists, fourth class ($12.00).
  Musicians, second class.
Specialists, fifth class ($8.00).
  Musicians, third class.

Men covered by this paragraph will be carried on muster rolls as privates, first class, or privates, with appropriate specialists ratings.

(18) *Headquarters and staff officers*.—All privates or privates, first class, detailed as clerks and messengers for duty at Headquarters Marine Corps, Headquarters Department of the Pacific, and in other staff offices within the United States, by the Major General Commandant or department commanders, will, when so detailed, thereby become entitled to the rating of specialist, third class, if detailed as clerks, and specialist, fourth class, if detailed as messengers.

(19) *Commanding officers will personally control the policy of rating men* of their commands as specialists within the authorized allowances, to the end that only men who are in fact competent specialists shall receive extra compensation. The extra compensation is not to be granted simply because men are detailed for special duties, but will be given only where the men are especially competent to perform the duties assigned.

(20) *First and second class specialists*.—In exceptional cases, where the Government would be manifestly benefited by the rating of a private or private, first class, as a specialist of the first or the second class, the Major General Commandant may authorize particular men to be so rated. In recommending such a rating for any man, the commanding officer will state the specific

reasons why he believes it would be to the advantage of the Government in the particular case.

(21) *New commands.*—Whenever a new command is organized, it will be assigned an allowance of specialists by the Major General Commandant, unless such command is included within the terms of this article.

DEPARTMENT OF THE PACIFIC.

35

(1) *Limits.*—The Department of the Pacific will include all posts, detachments, depots, offices, and other organizations of the Marine Corps (except the recruiting service and the detachments afloat) on the Pacific coast of the United States and those in the Hawaiian Islands and Guam.

(2) *Command.*—The departmental commander will, under the direction of the Major General Commandant, command all marines included in the department in so far as the command is not reserved by law or regulation to other authority.

(3) *The staff of the departmental commander* will consist of the assistant adjutant and inspector, the assistant quartermaster, and the assistant paymaster, stationed at San Francisco, Calif., and of such other officers as may be ordered to report to the departmental commander for duty on the staff.

(4) *Next in command.*—During the absence or disability of the departmental commander, or in case of a vacancy, the assistant adjutant and inspector on the staff will perform the duties of the office, unless or until another officer is ordered by the Major General Commandant to perform them.

(5) *Duties.*—The departmental commander will maintain the strength of the posts, detachments, offices, and other organizations included within the department, by transfer of officers and men, in accordance with the necessities of the service, and with the general and special instructions issued by the Major General Commandant, and will transfer enlisted men to the recruiting service, western and mountain divisions, to detachments afloat upon application of commanding officers to fill vacancies in authorized complements, and from detachments afloat upon application of commanding officers in cases of undesirables, and may effect mutual transfers of individuals between ships of the Pacific Fleet and shore stations in the department when applications are submitted through proper channels.

(6) *Promotion and reduction.*—The departmental commander will, within the department, appoint and reduce noncommissioned officers (except as provided in article 555, Navy Regulations) in accordance with the necessities of the service and pursuant to article 607, Marine Corps Manual, and the general and special instructions of the Major General Commandant.

(7) *Discharges.*—The departmental commander will, within the department, issue the necessary orders and prepare discharge certificates for the discharge of enlisted men (*a*) upon expiration of enlistment, (*b*) for undesirability, inaptitude, or unfitness, (*c*) for physical or mental disability (disability to be determined by a board of medical survey), or (*d*) in pursuance of a sentence of a court-martial, all in accordance with established policy and general and special instructions issued by the Major General Commandant. He will also award good-conduct insignia in accordance with article 1003.

BRIGADE AT SAN DIEGO, CALIF.

36

(1) *Advanced Base Force.*—The organized units of the Advanced Base Force, U. S. Marine Corps, which are stationed in the Department of the Pacific shall be attached to the U. S. Marine Corps brigade therein located.

(2) *Other organizations.*—While such brigade remains at the naval base, San Diego, Calif., there shall be included under the command of the commanding

general thereof all officers and men and all organizations of the Marine Corps attached to and serving at the naval base, San Diego, Calif.

(3) *The commanding general* of this brigade shall exercise, subject to superior authority, full military authority over the units of his command; but he shall not direct or be responsible for the administration of such units and detachments of his command as may be attached to any of the various naval administrative establishments of the naval base. He shall, however, be kept informed of the military training, the discipline, and condition of the Marine Corps units attached to such establishments, and shall make such inspections from time to time as may be necessary to keep him acquainted with the condition of all parts of his command.

(4) *Correspondence* shall be routed through the commanding officer of the establishment concerned.

(5) *Next in command.*—In the absence of the commanding general of this brigade from station or duty at the naval base, the senior line officer of the Marine Corps on duty at such base shall exercise military authority over officers and men and organizations of the Marine Corps attached to and serving thereat in the manner prescribed in paragraph (3) of this article for the commanding general, and while so in command he shall be addressed and sign himself as "Commanding Officer of Marines, Naval Base, San Diego, Calif."

RECORD OF EVENTS.

37

(1) *When kept.*—A record of events on expeditionary duty, advanced base duty, and in campaign, will be kept by each company and higher organization of the Marine Corps. Entries will be made daily, the day comprising the 24 hours covered by the date and should form a concise history of military operations.

(2) *Arrangement.*—Each day's record will commence with a march table, or statement of the operations or location of the organization, including an account of weather, roads, camp, health of command, etc., and a statement of the supply of ammunition and rations. This will be followed by a chronological record of events, including time and place of issue and receipt of orders and messages, with a copy or synopsis of contents.

(3) *Exactness.*—It is of special importance that the exact hour and place at which movements are begun and ended, and orders or important messages sent or received, be noted. After an engagement, the record of events will contain a report of losses and captures and will be accompanied by a sketch showing the positions of the command at the most important phases.

(4) *Each day's record* will be attested by the commander and with attached copies of orders and messages sent and received, will be forwarded daily to the next higher commander, who, as soon as practicable after the receipt thereof, will forward them to the Major General Commandant. Commanders of units not components of a higher command will forward their record of events direct to the Major General Commandant.

(5) *Suspension of record.*—Commanding officers of expeditions, advanced base forces, and marine forces in campaign are authorized to modify or suspend the preparation of the record of events if in their opinion the nature of the duty does not justify committing all or any part of the events to record, in which case the modification or suspension will be reported to the Major General Commandant.

(6) *An extra copy* of reports required of commanding marine officers in accordance with article 805, Navy Regulations, will be made and marked "For the Major General Commandant, United States Marine Corps."

(7) *In addition* to the above there will be rendered periodical operations reports and intelligence reports.

BIRTHDAY OF THE MARINE CORPS.

38

The following will be read to the command on the 10th of November of every year:

(1) On November 10, 1775, a Corps of Marines was created by a resolution of Continental Congress. Since that date many thousand men have borne the name Marine. In memory of them it is fitting that we who are marines should commemorate the birthday of our corps by calling to mind the glories of its long and illustrious history.

(2) The record of our corps is one which will bear comparison with that of the most famous military organizations in the world's history. During 90 of the 146 years of its existence the Marine Corps has been in action against the Nation's foes. From the Battle of Trenton to the Argonne, marines have won foremost honors in war, and in the long eras of tranquillity at home generation after generation of marines have grown gray in war in both hemispheres, and in every corner of the seven seas that our country and its citizens might enjoy peace and security.

(3) In every battle and skirmish since the birth of our corps marines have acquitted themselves with the greatest distinction, winning new honors on each occasion until the term marine has come to signify all that is highest in military efficiency and soldierly virtue.

(4) This high name of distinction and soldierly repute we who are marines to-day have received from those who preceded us in the corps. With it we also received from them the eternal spirit which has animated our corps from generation to generation and has been the distinguishing mark of the marines in every age. So long as that spirit continues to flourish marines will be found equal to every emergency in the future as they have been in the past, and the men of our Nation will regard us as worthy successors to the long line of illustrious men who have served as "Soldiers of the Sea" since the founding of the corps.

FLAGS.

39

The following flags, pennants, and guidons are authorized for use in the Marine Corps:

Flags:

Marine Corps standard, silk, with staff.
National colors, silk, with staff.
Regimental, silk, with staff.
Garrison, bunting.
Post, bunting.
Storm, bunting.
Recruiting, bunting, with staff (for automobiles).
Recruiting, blue, bunting.
Recruiting, red, bunting.
Automobile, major general, bunting, with staff.
Automobile, brigadier general, bunting, with staff.
Boat, major general, bunting, with staff.
Boat, brigadier general, bunting, with staff.
Hospital, field, bunting.
Quarantine, field hospital, bunting.
Sanitary, cordon, bunting, with staff.
Signal, red, 2-foot, bunting, with staff.
Signal, white, 2-foot, bunting, with staff.
Signal, red, 4-foot bunting, with staff.
Signal, white, 4-foot, bunting, with staff.
Semaphore, bunting, with staff.

Pennants:

Brigade, large, bunting.
Brigade, small, bunting.
Post commander, boat, bunting, with staff.
Quartermaster's supply depot and train, bunting.

Guidons:

Silk, with staff.
Ambulance and dressing station, bunting, with staff.

COMMAND OF MARINES ON ARMY TRANSPORTS.

40

When regularly organized units of the Marine Corps are embarked for transportation on an Army transport the commanding officer of such units will command all members of the Marine Corps on board junior in rank. When marines who are not regularly organized as a unit or units are so embarked and there is no regularly organized unit on board, the senior line officer of the Marine Corps junior to the commanding officer of troops of the Army, or the senior line noncommissioned officer in the absence of such a commissioned or warrant officer, will command the members of the Marine Corps junior in rank.

PISTOLS.

41

(1) ***Safe-keeping.***—All commanding officers shall, by the promulgation of a post order, designate a suitable place (or places) properly secured by locks or other safeguards, where automatic pistols in use by individual enlisted men shall be kept at all times when not in actual use.

(2) ***Checkage for loss.***—The value of any pistol lost through the failure of any person to keep his automatic pistol in the place designated by his commanding officer for the purpose will, upon due investigation and after responsibility has been fixed, be checked against the account of the enlisted man concerned.

(3) ***Instruction before arming.***—Men will not be armed with automatic pistols until after they have received a thorough course of pistol instruction and drill and have fired the Navy Pistol Course or the course prescribed in Pistol Marksmanship.

ECONOMY IN MOTOR TRANSPORTATION.

42

(1) The strictest economy in the use of motor transportation must be observed at all posts and stations of the Marine Corps.

(2) The use of motor transportation will be confined strictly to official purposes, and every effort will be made to limit the use thereof to absolutely necessary work.

(3) A large truck will not be used when a small truck will do, or a small truck when a motor cycle will do.

TRANSPORTATION OF RECREATION SUPPLIES.

43

Supplies and equipment turned over by the Bureau of Navigation, Sixth Division, intended for Marine Corps organizations, may, where no Government transportation is available, be received and transported on Government bills of lading, the transportation charges to be billed against the Marine Corps.

ASSISTANCE FOR EX-SERVICE MEN IN THEIR RELATIONS WITH THE VETERANS' BUREAU.

44

(1) It is the purpose of Marine Corps Headquarters, acting through its available personnel, to assist ex-service men in every possible way in securing contact with the Veterans' Bureau, thus enabling them without delay to renew or convert their insurance, to secure medical or dental treatment, hospitalization, or vocational training, or to present their claims for compensation.

(2) The obligation is imposed upon all post commanders and recruiting officers of the Marine Corps to aid their less fortunate comrades. Such officers will at once familiarize themselves with the orders and circulars relating to the Veterans' Bureau, in so far as these instructions refer to renewal or conversion of insurance, compensation, medical or dental treatment, hospitalization and vocational training, so that intelligent assistance and advice may be afforded ex-service men in regard to their relations with the bureau.

(3) The officers mentioned will obtain direct from the Veterans' Bureau a supply of all blank forms needed by ex-service men in their contact with that bureau.

(4) Upon application from ex-service men the officers mentioned in paragraph (2) will furnish blank forms, will assist in the preparation of applications, will carefully examine all papers or instructions on the forms in question, and will themselves promptly forward the completed applications or statements directly to the United States Veterans' Bureau, Washington, D. C.

(5) In cases where it is necessary to forward a copy of the discharge certificate of the ex-service man the copy will be made by the post commander or recruiting officer, adding the words "Copy of discharge certificate, to be used *only* with claims against the United States Veterans' Bureau," at the beginning of the copy, and the words "I hereby certify that the foregoing is a true, literal, and exact copy of the discharge certificate of ——— ———," with signature and rank of the attesting officer, at the end of the copy. Ordinary blank paper or the regular discharge certificate blanks (in the case of ex-marines) may be used in making these copies, except that parchment or artificial parchment shall not be used for this purpose. The attestation must be by a post commander, recruiting officer, notary public, or other person authorized to administer oaths.

CHAPTER 2.

ENTRY INTO THE SERVICE.

Section 1.—OFFICERS.

APPOINTMENT AS SECOND LIEUTENANT.

201

(1) *Officers authorized to recommend for.*—General and field officers and officers in command of companies and detachments of the Marine Corps are authorized to recommend, through official channels, meritorious noncommissioned officers for advancement to the grade of second lieutenant. In making such recommendations, the officer will be guided by the following instructions.

(2) *Age.*—In accordance with the law governing appointments to the grade of second lieutenant, a noncommissioned officer candidate must be between the ages of 21 and 27 when appointed.

(3) *Recommendations.*—The recommendations will contain specific statements as to the physical, mental, and moral fitness of the proposed candidate, and will be accompanied by available documentary evidence of character, experience, and personal history.

(4) Statements as to physical fitness will be based upon a medical examination, to determine if any physical defects exist, and upon observation, to determine if physical energy or the contrary is indicated. Statements as to mental fitness will be based upon obtainable evidence of scholastic education and upon observation of mental qualities indicated, such as mental energy, endurance, alertness, intelligence, adaptability, interest in the service, and their contraries. Statements as to moral fitness will be based upon evidence obtained from those in a position to have reliable knowledge of the moral development of the proposed candidate during the formative period of his life—during the ages from about 10 to 20 years—and upon personal observations and reports, official or otherwise, relating to the character and habits of the proposed candidate. His associates, language, deportment, and methods of amusement and recreation should be carefully noted and mentioned. His attitude in relation to self-discipline and willing obedience to orders, regulations, and law should be studied and reported upon.

(5) Recommendations received at Headquarters will be studied in connection with the military history of the proposed candidates and selections will be made by the Major General Commandant.

(6) *Preliminary examinations.*—When any proposed candidate has been selected by the Major General Commandant, a set of preliminary examination questions will be sent to the noncommissioned officer's commanding officer, who will have him take the examination under proper supervision. Upon completion of the examination, the papers thereof will be forwarded to the Major

General Commandant. After the results of the examination shall have been considered, the proposed candidate will be informed as to whether he has been selected as a candidate for a commission.

(7) *Selected candidates.*—Those noncommissioned officers selected as candidates will be transferred to the candidates' class at the Marine Barracks, Washington, where they will be under special observation and instruction preliminary to taking the final examinations for appointment to the grade of second lieutenant.

(8) During the period of instruction and observation the officers designated for that purpose will report from time to time relative to the indicated qualifications and disqualifications of the respective candidates. In case an opinion is formed that any candidate is in any respect unqualified for a commission, or in case a serious doubt is raised as to the positive fitness of any candidate, the Major General Commandant will direct his detachment and he will cease to hold the status of a candidate.

(9) *Withdrawal of nomination.*—An officer who has nominated any noncommissioned officer as a candidate may, for cause, withdraw his nomination at any time prior to the actual commissioning of such noncommissioned officer.

(10) *Scope of examination.*—The scope of the preliminary examination and of the final mental and professional examinations will be announced annually by the Major General Commandant.

(11) *Policy of Major General Commandant.*—It is the policy of the Major General Commandant to give full effect to the legislative provisions which open the way for advancement to the commissioned grades to meritorious enlisted men of the Marine Corps. At the same time, the mere fact of honest and faithful service in the ranks will not be accepted in lieu of other essential characteristics which are considered essential in one who is to serve as a commissioned officer in the Marine Corps. Keeping the door open, but selecting those who are to pass through, is one of the most important of the duties with which any officer may interest himself. While calling for specially intelligent thought and some additional labor, the officer who aids in making effective the general policy indicated will do much to assure the high professional standing of the near and the distant future Marine Corps.

Appointment of Marine Gunners, Quartermaster Clerks, and Pay Clerks.

202

(1) *Law and regulations.*—The act of August 29, 1916, provides that marine gunners and quartermaster clerks shall be appointed from the noncommissioned officers of the Marine Corps. Article 1645, Navy Regulations, provides that warrant officers and pay clerks of the Marine Corps shall be appointed in such manner as may be prescribed by law after their qualifications have been established in such manner as the Secretary of the Navy may prescribe.

(2) *Qualifications.*—The Secretary of the Navy has prescribed that the qualifications of candidates for appointment as marine gunner, quartermaster clerk, and pay clerk shall be established as follows:

"Men who have held temporary commissions, warrants, or appointments as pay clerk in the Marine Corps may, upon passing the necessary physical examination before a board of medical examiners, be appointed warrant officers or pay clerks if the Major General Commandant is satisfied from ther records that they are mentally, morally, and professionally qualified for appointment. All other candidates shall, in addition to the physical examination, be subject to such mental, moral, and professional examination as the Major General Commandant may prescribe."

(3) *Examinations.*—Candidates for appointment as marine gunner, quartermaster clerk, and pay clerk shall submit to boards appointed for their exami-

nation such evidence as may be required in respect to their mental and moral qualifications. In addition they shall be examined as follows:

For marine gunner, in the subjects of—
- Grammar and composition.
- Arithmetic.
- Infantry Drill Regulations, up to and including the platoon; organization and administration of the squad, platoon, and company; the nomenclature and functioning of infantry weapons (including machine guns adopted for use in the Marine Corps); visual signaling; knotting, splicing, and lashing; physical training; personal hygiene for the soldier; first aid.

For quartermaster clerk, in the subjects of—
- Grammar and composition.
- Geography.
- American history.
- Arithmetic.
- Regulations and orders.
- Administration, quartermaster's department.

For quartermaster clerk, adjutant and inspector's department, in the subjects of—
- Grammar and composition.
- Geography.
- American history.
- Arithmetic.
- Regulations and orders.
- Administration, adjutant and inspector's department.

For pay clerk, in the subjects of—
- Grammar and composition.
- Geography.
- American history.
- Arithmetic.
- Regulations and orders.
- Administration, paymaster's department.

General Policy and Instructions Concerning Appointment as Marine Gunner.

203

(1) Selections of noncommissioned officers for advancement to the rank of marine gunner will be based upon (1) the man's excellent moral character; (2) his excellent military character; (3) his knowledge and experience in Marine Corps activities.

(2) The development of military character, particularly those traits which inspire respect and confidence in both seniors and juniors, generally requires considerable service, as does the acquiring of experience. Knowledge may be acquired in a comparatively short time if a man is intelligent and has a fair basic education.

(3) A great deal of knowledge that would be most useful to a marine gunner might be acquired before he enters military service; but such knowledge is a secondary consideration to the military characteristics of a successful marine gunner.

(4) One of the purposes in establishing the rank of marine gunner was to offer a respectable and valuable reward to those who give long years of valuable service as enlisted men in the performance of line duties—the duties that carry them aboard ship, to foreign countries, to hardships of expedition and campaign, as well as the more or less routine duties of post and barracks.

(5) The general policy of the Major General Commandant is to fill vacancies by selections made on the foregoing basis. Ambitious self-improvement in the military profession will be recognized, as will exceptionally meritorious conduct in action. At the same time it will be an exceptional case where a noncommissioned officer will be advanced to the rank of marine gunner who has less than 10 years' service, or less than two discharges with "Character excellent."

(6) General and field officers, and post, company, and detachment commanders are authorized to submit recommendations for the advancement of noncommissioned officers to the rank of marine gunner at any time, keeping in mind the above general policy of the Major General Commandant in the matter. Such recommendations should state in detail the general and special qualifications of the proposed candidate; his military knowledge and experience in command, as well as his moral and military character, as disclosed to the officer submitting the recommendation.

(7) A file of such recommendations will be kept at Headquarters and will be consulted as vacancies are to be filled from time to time.

Section 2.—ENLISTED MEN.

204

ENLISTMENT.

(1) *The place of acceptance* of an applicant for enlistment is the place at which he applied for enlistment, was examined by the doctor, and was accepted by the recruiting officer or noncommissioned officer in charge.

(2) Where an applicant is examined and passed by the doctor and accepted at a station where there is no recruiting officer, the words "Examined and accepted at ———, 192—," will be inserted in the margin of the enlistment paper, N. M. C. 321 or N. M. C. 321B, and signed by the noncommissioned officer in charge.

205

Out
IC. 1.

(1) *Bona fide home or residence.*—Each man on enlisting or reenlisting must certify over his signature on both the enlistment paper and service-record book as to his bona fide home or residence, including town, county, and State. The entry on the enlistment paper, N. M. C. 321, may be accomplished by inserting in the declaration, after the words, "and that I am a citizen of the United States," or other convenient place, the words "and that my bona fide home or residence is ———." The entry in the service-record book will be made on page 1 by writing or stamping in the place provided for former residence "Bona fide home or residence ———."

(2) The bona fide home or residence is required in connection with the payment of travel allowance on discharge and for the purpose of compiling statistical data for crediting enlistments to the several States.

(3) At the time the certificate is furnished by the enlisted man he will be advised of its purpose, and informed that thereafter no change will be made therein except on approval of the Major General Commandant.

EXTENSION OF ENLISTMENT.

206

(1) *How and when made.*—The term of enlistment of any man may, by his written voluntary agreement, be extended for a period of either one, two, three, or four full years from the date of expiration of the then existing term of enlistment. An agreement to extend an enlistment must be executed prior to or at expiration of original enlistment; and a man serving an extension of

less than four years may, before expiration of such extension, further extend his term repeatedly by one or more full years, the aggregate of all extensions not to exceed four full years from the date of expiration of the original term; but no man shall be permitted to extend or reextend his term of enlistment where his retention in the service is not desirable, and at any time before an extension term begins to run the commanding officer may cancel the extension agreement should the man's conduct so warrant. However, unless specially authorized by the Major General Commandant, extensions for one year will be accepted only in the case of—

(*a*) A man on foreign or sea service, for the purpose of continuing thereon after expiration of term of enlistment.

(*b*) A man on shore duty within the continental limits of the United States who has applied for sea or foreign service, for the purpose of extending his enlistment period sufficiently to meet requirements for transfer for such service.

(*c*) A man on recruiting duty, or on duty at Headquarters Marine Corps or at detached staff offices or depots.

(2) *Supplemental to original contract of enlistment.*—The voluntary agreement to extend a term of enlistment shall be supplemental to the original contract of enlistment and form a part of it, and shall be executed in the terms and on the blank forms prescribed by the Major General Commandant.

(3) *Forwarding.*—When an agreement to extend enlistment has been completed it shall be immediately forwarded to the Adjutant and Inspector, Marine Corps, and entries of the extension made in red ink in the man's service-record book, over the signature of the commanding officer, at the bottom of page 1, and also on the line below the last markings on pages 4 to 9. Similar entries shall be made for any subsequent extensions, noting in addition the fact of its being a second, third, or fourth extension.

(4) *The records of men extending enlistments* shall be kept continuously in the service-record books of the enlistments extended, additional pages being inserted when necessary.

(5) *Upon the first extension of an enlistment* a transcript of the service markings, obtained as in closing the service-record book for discharge, a statement of the offenses committed during the enlistment, and appropriate recommendation as to character and good conduct medal as in case of discharge, will be entered on the written agreement of extension. In case the extension is accomplished before the day that the book and accounts will ordinarily be closed for discharge, the transcript, statement, and recommendation will be omitted; but the man's commanding officer will on that day forward to the officer authorized to award good-conduct medals a similar transcript, statement, and recommendation covering the entire period of the original enlistment. The good-conduct medal will not be awarded until the time that the discharge would ordinarily be issued.

(6) *Physical examination.*—A man desiring to extend his enlistment shall be required to pass the same physical examination as is required for reenlistment. and the examining surgeon's certificate shall be attached to the completed "Agreement to extend enlistment."

(7) *Who may extend.*—Commanding officers are authorized to accept the extension of enlistments of men who, in their opinion, are desirable for retention in the service and who would ordinarily be recommended for reenlistment. Such cases require no reference to the Major General Commandant, the agreement to extend enlistment being forwarded to the Adjutant and Inspector immediately upon its completion.

(8) *When executed.*—Generally, a man shall not be allowed to extend his enlistment until about the completion of his original term of enlistment, in order that it may be more satisfactorily determined whether he is specially desirable, what markings and character should be given him, and whether or not he should be recommended for a good-conduct medal.

FURLOUGHS.

207

Commanding officers of posts and recruiting officers are authorized to grant furloughs to men upon reenlistment as follows:

Upon reenlistment for three years, a furlough of two months.

Upon reenlistment for four years, a furlough of three months.

MEN WITH DEPENDENTS.

208

Recruiting officers will inform all men upon enlistment as follows:

(1) *Quarters.*—That there are no accommodations at Marine Corps posts for families or dependents of enlisted men, except a few for the highest noncommissioned officers.

(2) *Support.*—That an enlisted man whose permanent rank is not higher than sergeant can not properly expect to support any dependents on his pay and allowances.

(3) *Discharge.*—That requests for discharge based on dependency of wife or child will not be considered in cases in which marriage has taken place since enlistment, unless the marriage has received the prior approval of the Major General Commandant.

RECRUITING INSTRUCTIONS.

209

Complete instructions as to physical and other requirements for entry into the service as enlisted men will be found in "Recruiting Instructions, U. S. Marine Corps."

CHAPTER 3.

CLOTHING AND EQUIPMENT.

Section 1.—OFFICERS.

301

Sale of clothing, etc., to officers.—The depot quartermaster, Philadelphia, is authorized to sell uniforms, accouterments, and equipment at cost price to officers of the Marine Corps. Officers who have access to the Depot of Supplies, Philadelphia, may have uniforms manufactured at that depot. The depot quartermaster is, however, authorized to refuse to manufacture clothing for officers except in those cases where two or more "try-ons" can be made, unless the officer concerned agrees to accept the clothing as manufactured without such "try-ons," in which latter event the officer concerned will be required to make payment for the clothing manufactured whether or not the fit proves satisfactory. Those officers who are not in position to visit the depot may purchase the necessary uniform cloth, accouterments and equipment from the depot quartermaster, Philadelphia, or from the depot quartermaster, San Francisco. Applications should be made direct to the depot quartermasters mentioned for such articles as are required, or for information in reference thereto.

Section 2.—ENLISTED MEN.

304

Retention by Government, men dishonorably discharged, etc.—When a man is dishonorably discharged, or is discharged with a bad conduct discharge pursuant to the sentence of a summary court-martial on conviction of theft or other offense involving great moral turpitude, the outer uniform clothing in his possession will be retained by the Government.

ISSUE OF EQUIPMENT.

305

(1) *To recruits.*—All men under instruction at recruit depots, upon going on the range, or upon transfer before going on the range, shall be issued the following equipment:

(*a*) Rifle, United States, caliber .30.
Bayonet for.
Brush and thong for.
Cover, front sight, for.
Case, oiler, and thong for.
Scabbard, bayonet, for.
Sling for.

(b) One Infantry equipment complete, consisting of the following:
1 belt, cartridge, rifle, model 1910.
1 canteen, model 1910.
1 can, bacon.
1 can, condiment.
1 can, meat.
1 carrier, pack (haversack).
1 cover, canteen.
1 cup, canteen.
1 haversack.
1 knife, haversack.
1 fork, haversack.
1 spoon, haversack.
1 poncho.
1 package, first-aid.
1 pouch, first-aid.
1 pouch, meat can (haversack).

(2) *To others.*—The above-mentioned equipment will be issued to all other enlisted men of the Marine Corps, except men on duty at Headquarters Marine Corps or one of the staff offices or on recruiting duty, and members of the Marine Band. Sergeants major, quartermaster sergeants, first sergeants, gunnery sergeants, and field musicians will have permanently issued to them only the equipment enumerated under paragraph (1) (b) and the belt, pistol, will be substituted for the cartridge belt, rifle.

(3) *Disposition of equipment.*—All men having once been issued this equipment will retain it in their possession until discharged, except that men transferred from ships outside the continental limits of the United States and foreign shore station to naval hospitals in the United States, upon report of medical survey, and men transferred to the Army and Navy General Hospital, Hot Springs, Ark., the Fitzsimmons General Hospital, Denver, Colo., or to St. Elizabeths Hospital, Washington, and men transferred to duty at Headquarters Marine Corps or one of the staff offices, or to recruiting duty, will be required to turn in the equipment at time of transfer. In the case of men on board ship in the United States transferred to a naval hospital, the equipment will be invoiced and sent to the post quartermaster at the marine barracks to which the man's staff returns are sent, the invoice being marked with the man's name, so that the equipment may be reissued to him on reporting for duty.

(4) *Rifle used at target practice.*—Each man will be required to use his own rifle at target practice. Any rifle the shooting qualities of which are poor will be exchanged for a good rifle.

(5) *Replacement.*—This article will not prevent replacing of any part of equipment which may become unserviceable.

(6) *Inspection.*—Officers are cautioned as to the necessity of frequently inspecting the arms and equipment of men whose duties do not provide for their presence at the usual routine inspections.

(7) *Rifle, care of.*—Bores of rifles will habitually be kept coated with a thin coating of light oil. The use of emery powder or any other polishing material in cleaning the rifle is forbidden.

(8). *Rifle, number of.*—The number of the rifle issued each man will be entered in his service-record book.

(9) *Price list.*—The price list of clothing, equipment, etc., is published yearly in Marine Corps Orders.

Issue of Clothing.

306

(1) *Recruits.*—In order to prevent hardships being imposed upon men entering the Marine Corps, only such quantities and articles of clothing as are listed below will be issued upon enlistment or reenlistment:

1 belt, trousers, web.
2 blankets, woolen.
2 coats, service, summer.
1 coat, service, winter.
4 drawers, knee or woolen.
1 gloves, woolen, pair.
1 hat, field.
2 leggings, pairs.
1 ornament, bronze.
1 ornaments, collar, bronze, set.
1 overcoat, service, winter.
2 shirts, flannel.
4 shirts, under, cotton or woolen.
2 shoes, russet ~~or cordovan~~, pairs.
4 socks, cotton or woolen, pairs.
3 trousers, service, summer.
1 trousers, service, winter.

Woolen underwear, gloves, and socks will only be issued when the season so warrants.

Hat letters may be issued where required.

(2) The above should be sufficient to last for three months, and no additional clothing will be issued to a man during the first three months of his enlistment except in case of an emergency, such as extreme weather requiring an additional blanket, and then only with the approval of his commanding officer, or in case a man is transferred to tropical or sea service.

307

General court-martial prisoners.—Commanding officers authorizing the issue of necessary clothing to marines sentenced to forfeiture of pay and allowances will name the articles so authorized in a written order on the issuing or accountable officer, who will place on each issue slip a notation reading as follows:

"This clothing issued as necessary for the health and comfort of the prisoner named hereon by order of

Name_______________________________________

Rank, USN or USMC________________

Letter dated__________________________."

CHAPTER 4.

PAY AND ALLOWANCES.

401

Paid monthly.—Enlisted men of the Marine Corps serving at navy yards or barracks shall be paid monthly.

402

(1) *Entry of pay data in service-record books.*—Commanding officers of marines ashore and afloat charged with the keeping of the service-record book of any marine will immediately after the payment of each monthly pay roll cause to be entered and verified without delay, in the space provided in the service-record book, a record of the payment or settlement as shown by the pay roll as audited and settled by the paymaster concerned, entering balance "Overpaid" or "Unpaid" if any; and if none, entering by means of stamp the words "paid in full" through the space provided for the entering of such balances.

(2) At the time of transfer of an enlisted man whose pay account records, as shown in the service-record book, stands overpaid or due and unpaid at date of last settlement, care will be taken to see that such balances agree with the pay roll upon which such settlement appears. If correct, the entry (in figures only) will be initialed in the opposite space by the commanding officer having charge of the record. If not correct, proper correction will be made in the figures and immediately on the line following an entry will be made thus: "Corrected to read—Overpaid ——— dollars," or "Corrected to read—Due and unpaid ——— dollars," as the case may be, writing such amount in words, this entry to be signed by the commanding officer concerned.

403

Closing accounts.—To enable the pay officers of ships or stations, carrying the rolls of enlisted men, to transfer the accounts of dead men, deserters, and general court-martial prisoners to the General Accounting Office, Navy Department Division, Washington, D. C., to the deserters' roll at Headquarters, U. S. Marine Corps, and to naval prisons, respectively, as well as to enable them to prepare final settlements preliminary to discharge, commanding officers of marines will furnish such pay officers with a detailed statement of account in all such cases, prepared on form N. M. C. 90, revised.

TRAVEL PAY.

404

(1) *Travel Pay.*—When an enlisted man who enlisted or reenlisted prior to August 1, 1921, is about to be discharged, the commanding officer will take the man's affidavit as to his bona fide home, or residence, and will enter the follow-

ing information under "Remarks" on form N. M. C. 90 (Statement closing account for settlement), in his case, or will attach the same thereto:

"Actual bona fide home or residence as sworn to this date ——— ———.

"'Home post office' or address as shown on Form 1-B attached to service-record book ———. Claimant elects travel pay to the more distant point." (If such be the case.)

(2) Affidavits taken in accordance with instructions contained herein will be permanently attached to service-record books.

(3) Paragraphs 1 and 2 of this article do not apply to members of the Marine Corps Reserve, since information of the actual bona fide home or residence in their cases is on file at Headquarters, and is furnished in orders for disenrollment or to inactive status.

(4) When an enlisted man who enlisted or reenlisted after July 31, 1921, is about to be discharged under conditions entitling him to travel allowance, the commanding officer in submitting form N. M. C. 90 (statement closing account for settlement) to the pay officer concerned, will enter under "Remarks," the following information, or will attach the same thereto:

"Enlisted at ———(Place). Actual bona fide home or residence as certified at date of enlistment. ———(Place), (or as changed and approved by the Major General Commandant, ———(Date)

"Claimant elects travel allowance to the more distant point." (If such be the case.)

405

The enlisted grades of the Marine Corps are grouped, in order to classify them with their corresponding grades of the Infantry of the Army, for pay purposes, as follows:

- First grade (master sergeants, Army):
  - Sergeants major, Marine Corps.
  - Quartermaster sergeants, Marine Corps.
- Second grade (first sergeants, Army):
  - First sergeants, Marine Corps.
  - Gunnery sergeants, Marine Corps.
  - Drum majors, Marine Corps.
- Third grade (staff sergeants, Army):
  - (None) Marine Corps.
- Fourth grade (sergeants, Army):
  - Sergeants, Marine Corps.
- Fifth grade (corporals, Army):
  - Corporals, Marine Corps.
- Sixth grade (privates, first class, Army):
  - Privates, first class, Marine Corps.
- Seventh grade (privates, Army):
  - Drummers, Marine Corps.
  - Trumpeters, Marine Corps.
  - Privates, Marine Corps.

Extra Compensation.

406

(1) *Signalmen.*—Enlisted men of the Marine Corps regularly detailed as signalmen are entitled to the same extra compensation in addition to pay as is now or may hereafter be allowed enlisted men of the Navy.

(2) *Cooks, etc.*—The following extra compensation, which shall be credited on the monthly pay rolls, is allowed enlisted men of the Marine Corps, under regu-

larly authorized assignments or details, for service with messes composed of enlisted men of the corps ashore or afloat, viz:

	Per month.
First-class cook	10.00
Second-class cook	8.00
Third-class cook	7.00
Fourth-class cook	5.00
Messman	5.00

409

Longevity pay affected by unauthorized absence.—Enlisted men will be denied credit for periods of unauthorized absence involving checkages of pay in fixing the dates which will control the rate of pay described by the act of June 4, 1920, and for longevity increase. The original dates in each case will be extended by an amount equal to the aggregate of unauthorized absences for which pay has been checked.

410

Detailed instructions regarding pay and allowances will be found in the Paymaster's Manual and the System of Accountability.

CHAPTER 5.

OPERATIONS AND TRAINING.

Section 1.—GENERAL.

TRAINING OF OFFICERS, STAFF DETAILS.

501

(1) Infantry is the foundation on which the military structure is builded. It is the first duty of every Marine officer, therefore, to make himself a good infantryman. This does not mean that it is necessary for officers to devote all of their time to the study of infantry tactics. During the years of peace officers have at their disposal ample time in which to become proficient in all branches of the profession of arms.

(2) Especially is it essential that officers should familiarize themselves with the duties of the military and administrative staffs. An efficient staff is of vital importance to the successful functioning of a military organization. Under the detail system, all officers of the Marine Corps are eligible for temporary assignment to any of the staff departments in exactly the same manner as they are eligible for assignment to any other duty which the Marine Corps is required to perform. Their records as Marine officers, therefore, depend on the efficiency shown by them in the performance of both line and staff duties.

(3) In conformity with the principles enunciated in the preceding paragraphs, the following policy will be followed, viz:

(*a*) All officers will be prepared for assignment to either line or staff duty.

(*b*) Approximately one-fourth of the officers detailed in each staff department will be relieved therefrom during each calendar year.

(*c*) Officers, after relief from staff details, will not be detailed in a staff department for a period of at least two years.

(*d*) Officers below the grade of field officer, after relief from a detail in a staff department, will perform straight line duty with troops for a period of two years. This precludes their assignment during this period to duty as adjutant, acting assistant quartermaster, acting assistant paymaster, inspector, post-exchange officer, recruiting officer, aid-de-camp, permanent judge advocate of a general court-martial, or duty at Headquarters Marine Corps, or at any headquarters or staff office.

(*e*) No officer below the rank of captain will be regularly detailed as a member of a staff department. First lieutenants are eligible for assignment for duty as adjutant, acting assistant quartermaster, acting assistant paymaster, post exchange officer, recruiting officer, aide-de-camp, permanent judge advocate of a general court-martial, or to duty at Headquarters U. S. Marine Corps, or at any headquarters or staff office. Second lieutenants will not be assigned to any of the above-mentioned duties, but will be required to perform line duty with troops.

(4) Officers detailed in a staff department should give to the performance of staff duty the same initiative, industry, and energy as they give to the performance of line duty.

Broadly speaking, there is in the Marine Corps no separation of the line and staff. All are Marine officers, and all duties are Marine Corps duties.

REGULATIONS ARMY, NAVY, WHEN GOVERNING.

502

The exercises and formation of marines at parades, reviews, inspections, escort, guard mount, funerals, and the regulations prescribing salutes, shall be the same as those prescribed for the Navy (see Landing Force Manual). Other than as prescribed in the foregoing, the Infantry Drill Regulations, the Small Arms Firing Regulations and the Bayonet Training Regulations for the Marine Corps shall be the same as those prescribed for the Army. Duties of sentinels and internal regulations for camp and garrison duties shall be similar to those prescribed for the Army.

DUTIES OF OFFICER OF THE DAY.

503

(1) The duties of the officer of the day shall be conducted in accordance with instructions and regulations established for the Army.

(2) He shall visit the guards and the sentinels at such times, during his tour of duty, as may be prescribed.

(3) He shall attend all roll calls, and at posts where there are no organized companies shall inspect the men at all mess formations.

(4) At the hour designated by his commanding officer, he shall thoroughly inspect the grounds, quarters, bakehouse, mess room, cells, and sinks.

(5) In case of fire at the station, he shall immediately have the alarm sounded in the prescribed way, and inform the commandant of the station and his commanding officer and carry out the fire regulations of the station.

(6) He shall enter each day, in the sergeant of the guard's report book, the hours he visited sentries, and such other data as may be directed by his commanding officer.

(7) He shall carefully examine the sergeant of the guard's report book before guard mounting, and have any errors therein corrected. After having assured himself as to its correctness and completeness, he shall make the following entry over his own signature: "I have personally examined this report and find it to be correct," and will present it at office hours to the commanding officer for his inspection.

FIELD EQUIPMENT.

504

(1) *Officers.*—The following is the field equipment prescribed for officers of all ranks up to and including the rank of colonel:

- (*a*) Automatic pistol, .45 caliber, 3 magazines, holster, lanyard, and 21 rounds of ammunition.
- (*b*) Pistol belt, with suspenders and pouch.
- (*c*) First aid package and pouch.
- (*d*) Canteen, with cover and cup.
- (*e*) N. C. staff haversack, with meat can, knife, fork, and spoon.
- (*f*) Field glasses.
- (*g*) Whistle.
- (*h*) Poncho.
- (*i*) Flashlight, electric.
- (*j*) Wrist watch, with illuminated dial.
- (*k*) Compass, with illuminated dial.
- (*l*) Dispatch case, notebook, black, blue, and red pencils.
- (*m*) Clothing roll.
- (*n*) Bedding roll.
- (*o*) Trunk locker. (Field officers, 2 trunk lockers.)

(2) Articles (*a*) to (*h*) may be obtained from the Quartermaster's Department upon memorandum receipt; articles (*i*) to (*o*) may be purchased from the Quartermaster's Department.

(3) *Enlisted men; for foreign tropical expeditionary duty.*—When orders are received to transfer detachments to foreign tropical expeditionary service, it shall be the duty of the commanding officer, company, or detachment commander to personally satisfy himself that each man to be so transferred is supplied, prior to transfer, with all articles of equipment specified in article 305, and that each article is in good serviceable condition, issues being made to replace worn-out or damaged articles. He will also see that each man is supplied with a clothing bag, and the necessary toilet articles, including toilet and washing soap.

(4) Companies and detachments will leave their posts in complete field uniform and fully equipped, except that shelter-halves will not be taken unless specifically directed. Overcoats will always be taken, being worn when the weather requires. While abroad, especially in the tropics, overcoats will accompany the men wherever they are serving, this being required in order that they may have overcoats when ordered to the United States.

(5) Officers commanding detachments in the tropics will, before the men are transferred to the United States, see that they are properly supplied with a sufficient number of flannel shirts to assure their warmth and comfort upon their arrival in the United States during cold weather.

(6) The men will carry with them in their clothing bags, so far as may be practicable, all their clothing and personal effects. (See par. 213, System of Accountability.)

(7) Should any clothing be left at posts, it will be packed in accordance with instructions contained in the System of Accountability.

(8) Each company will take its company typewriter, desk, and four automatic pistols, one for each field musician and one each for the first sergeant and gunnery sergeant.

THE ADVANCED BASE FORCE.

505

(1) Marine Corps organizations available for overseas service with the fleet shall be known as "The Advanced Base Force, U. S. Marine Corps."

(2) Permanently organized units of the Advanced Base Force will normally be known by their numerical designation. Any part of the Advanced Base Force assigned to a special duty (force of occupation, landing forces, raiding forces, etc.) will be named in accordance with its mission.

(3) The training centers of the Advanced Base Force will be at Marine Barracks, Quantico, Va., and Marine Barracks, San Diego, Calif.

(4) The Advanced Base Force, or any part thereof, shall comprise such infantry, artillery, and specialist troops as may be assigned to it by Headquarters, U. S. Marine Corps, and shall receive instruction in the following branches and such other kindred subjects as may be designated by Headquarters, U. S. Marine Corps, to enable it to operate efficiently on any duty which may be assigned:

Artillery: Field guns, naval guns, howitzers, mortars, antiaircraft armament.
Fire control.
Machine guns, ground and antiaircraft.
Searchlights: Harbor defense, antiaircraft, and field.
Signals: Radio, telephone, and telegraph, visual, radio director, microphone sound detector.
Engineering.
Infantry and attached weapons, grenades, 37 mm., Stokes mortar, etc.
Air forces: Land and water planes, observation balloons.

MILITARY SCHOOLS.

506

(1) *Policy.*—The following is the policy of the Marine Corps in connection with the military schooling of commissioned officers. All officers mediately or immediately connected with the administration of such schooling will cooperate in the endeavor to obtain the desired results.

(2) The principal agency for such military schooling will be the Marine Corps schools at Quantico, Va.

(3) *Courses.*—The Marine Corps schools will conduct three regular courses for commissioned officers; one, the basic course; two, the company officers' course; and, three, the field officers' course.

(4) *Basic course.*—All officers appointed second lieutenants will be given the basic course as soon as practicable after appointment.

(5) *Company officers' course.*—All lieutenants will be given the company officers' course when practicable anterior to becoming due for promotion to the grade of captain, except those who shall have had the benefit of equivalent schooling theretofore. In addition, all captains who have not had the benefit of equivalent cchooling will be gven the campany officers' course when the exigencies of the service permit.

(6) *Field officers' course.*—All captains will be given the field officers' course when practicable anterior to becoming due for promotion to the grade of major, except those who shall have had the benefit of equivalent schooling theretofore. In addition, all lieutenant colonels and majors who have not had the benefit of equivalent schooling will be given the company officers' course when the exigencies of the service permit.

(7) *Infantry School, Camp Benning, Ga.*—In lieu of taking the company officers' course of the Marine Corps schools, a limited number of captains and senior first lieutenants who are graduates of the basic course or have had equivalent schooling may be assigned to take the company officers' course at the Infantry School, Camp Benning, Ga. Similarly a limited number of majors and senior captains who are graduates of the company officers' course or have had equivalent schooling may be assigned to take the field officers' course at the Infantry School.

(8) *School of the Line, General Staff School, Army General Staff College.*—A limited number of the field officer graduates of the field officers' course of either the Marine Corps schools or the Infantry School may be assigned to take the course at the School of the Line, Fort Leavenworth, Kans., and from graduates of that school there may be selected a limited number to take the course at the General Staff School at Fort Leavenworth. From graduates of the latter school a limited number may be selected to take the course at the Army General Staff College, Washington.

(9) *Naval War College.*—A limited number of officers above the grade of captain who are graduates of one of the prescribed field officers' courses or have had equivalent schooling may be assigned to take the course at the Naval War College, Newport, R. I.

(10) *Army Artillery schools.*—A limited number of officers may be assigned to take the battery officers' or the field officers' course at the artillery schools of the Army. Such officers will be selected from graduates of the company officers' course, Marine Corps schools, or from field officers, captain, and senior first lieutenants with equivalent schooling, depending upon the artillery course which they are to take.

(11) *Chemical Warfare School, Signal Corps School, Army.*—A limited number of officers may be assigned to take courses in other special service schools of the Army where courses are open to officers of the Marine Corps by courtesy of the Army, such as the Chemical Warfare School and the Signal Corps School. Such officers will be selected from senior first lieutenants and captains who are grad-

uates of the company officers' course, Marine Corps schools, or who have equivalent schooling.

(12) *Other military schools, Naval Aviation School.*—Assignments to other military schooling which may be available for commissioned officers of the Marine Corps as, for instance, Naval Aviation School, will be confined to those who shall have had the requisite schooling in infantry subjects for their respective ranks as herein indicated. In every such case the beneficial results to the service to be expected will be the determining factor.

(13) *Office of Judge Advocate General.*—In connection with the last preceding statement, junior officers with comparatively short commissioned service will not be assigned to duty in the office of the Judge Advocate General of the Navy.

(14) *Object.*—The particular object of the policy in question is to assure and hasten the military schooling of the commissioned personnel as a whole, as opposed to the more comprehensive schooling of a small minority, to assure to each officer basic schooling according to his rank, in connection with the main requirement that marines be excellent infantry men trained as well to serve at sea. At the same time it is not proposed to overlook opportunities for additional schooling whereby officers may not only specially educate themselves in particular military subjects but, also, extend their general military education.

Marine Corps Institute.

507

(1) *Enrollment.*—Officers and enlisted men desiring to enroll as students in the Marine Corps Institute, or wishing information relative to the courses offered, should communicate through their commanding officer direct with the director, Marine Corps Institute, Marine Barracks, Washington, D. C. Upon the receipt of inquiries or requests for enrollments, the director of the institute will forward all information and necessary enrollment papers to the applicant via his commanding officer.

(2) *Correspondence.*—After enrollment has been accomplished, correspondence with regard to the course will be carried on direct between the director of the Marine Corps Institute and the student.

(3) *Diploma.*—When an enlisted man satisfactorily completes a course in the Marine Corps Institute, a diploma or certificate will be awarded and transmitted to the student via his commanding officer, by the director. The commanding officer will make an appropriate entry in the service record of the man concerned, showing course satisfactorily completed and date of completion. The presentation of a diploma should be an occasion of ceremony.

(4) *Transfers.*—Upon the transfer of an enrolled student of the Marine Corps Institute, his immediate commanding officer will inform the director, Marine Corps Institute, Marine Barracks, Washington, D. C., of the fact, giving name of station or ship to which transferred, and date. A report should be made also in case of death, desertion, or discharge.

Section 2.—TARGET PRACTICE.

Courses Adopted.

508

(1) *The course in small-arms target practice* prescribed by War Department Document No. 1021, entitled "Rifle Marksmanship," as approved by the Secretary of War on June 23, 1920, has been adopted for use by the Marine Corps.

(2) *The course in pistol practice* prescribed by Parts I, II, IV, War Department Document No. 1050, entitled "Pistol Marksmanship," approved by the War Department on November 26, 1920, has been adopted for use by the Marine

Corps and will be fired in addition to the pistol course as outlined in Small-Arms Firing Regulations, U. S. Navy, by all officers and men whose duties may require skill in pistol marksmanship.

(3) *The course outlined in "Automatic Rifle Marksmanship,"* War Department Document No. 1011, as approved by the War Department on April 16, 1920, has been adopted for use by the Marine Corps.

509

Preliminary instruction.—Steps will be taken by commanding officers to have all men thoroughly instructed in the preliminary drills and in part 1 of the Marine Corps score book and in Rifle Marksmanship, and wherever practicable an officer will be detailed to supervise this instruction. It has been found that men from posts where attention is given to the preliminary drills qualify in large numbers, while the percentage of men qualifying from posts where this important work is neglected is small. All the preliminary drills, including instruction in part 1 of the score book and Rifle Marksmanship and gallery practice, simulating range firing in all its details to the greatest possible extent, can be carried on during the winter months with profitable results.

510

Insignia.—A man will be furnished appropriate insignia for his first qualification as expert rifleman, sharpshooter, or marksman. When a man qualifies three times (not necessarily successive) as expert rifleman or as sharpshooter, he will be awarded an expert rifleman's or sharpshooter's bar as may be appropriate. This provision also applies to officers.

511

(1) *Record course to be fired each year.*—It is the desire of Headquarters U. S. Marine Corps that every man fire the record course each year, and commanding officers will make every effort to carry out this desire. Each commanding officer, upon receiving the information that the rifle range to which he is required to send the men of his command for qualification is open, is authorized to communicate with this range direct and to arrange for the transfer of such members of his command as are available for target practice. The detachments sent from each post should be of such size as will enable the men of the command to complete the target firing about one month before the probable closing date for the rifle range. The last month will be used to fire such men as have joined during the season and such other men as have not been able to fire before that time.

(2) *Transportation.*—Where ranges are located at points distant from stations commanding officers, for the purpose of holding regular practice under Rifle Marksmanship or under Firing Regulations for Small Arms, United States Navy, are hereby authorized to direct post quartermasters to furnish the necessary transportation to officers and enlisted men, and to order the travel necessary. For the purposes mentioned, officers will generally perform travel with troops, and no orders for the above purposes issued by commanding officers to officers will entitle them to mileage unless orders are approved by the Major General Commandant.

QUALIFICATIONS.

512

(1) *Extra pay.*—Enlisted men of the Marine Corps who are now qualified or who may hereafter qualify as expert riflemen, sharpshooters, or marksmen under tests in all respects the same as those authorized for the Army shall

receive the same extra pay under the same conditions as may now or hereafter be provided for enlisted men of the Army by Army Regulations.

(2) *Marine Band.*—Qualifications can not be made in the Marine Band, nor will extra compensation for markmanship be paid to enlisted men in the Marine Band.

(3) *Eligibility for qualification.*—All other enlisted men in the Marine Corps are eligible for qualification and entitled to extra compensation therefor.

(4) *Credit on pay rolls.*—Immediately upon qualification under the Army course men will be credited upon the next succeeding pay roll with the extra compensation from the date of qualification, and the following notation will be made upon the first pay roll upon which credit is made: Qualified as marksman (sharpshooter or expert rifleman), (date), per certification by (name of officer certifying target record).

(5) *Publication of qualification.*—The fact of qualification will be published in orders from Headquarters, United States Marine Corps, and upon receipt of such order the date and number of the order will be noted in the service record book of each man concerned.

513

Original record.—The original record of marksmanship qualification is the entry in the service-record book. Officers signing pay rolls are responsible for the entry of necessary data in regard to qualifications on the pay rolls, and such officers are cautioned as to the necessity for using care to prevent overpayment when qualification has ceased under the regulations, particularly when preparing N. M. C. 90 in closing accounts for discharge.

514

Qualifications, duration of.—All qualifications announced in these orders begin on the date of qualification as noted in the order and continue until the next opportunity to requalify or for one year if no opportunity for requalification is presented within that period, provided that, if discharged during that time, the man reenlists within three months from the date of discharge.

515

Reenlistment.—When a man who is entitled to extra compensation for marksmanship at the time of his discharge reenlists within three months from the date of said discharge, his qualification continues and he is entitled to the extra compensation therefor until the first opportunity to fire subsequent to reenlistment, or for one year from the date of his last qualification if no opportunity to fire for qualification is presented within that period.

516

Requalification.—When men holding qualifications in marksmanship increase their qualifications, the new qualifications become effective on the dates on which made. This applies also to qualifications equal to those already held. When men holding qualifications in marksmanship fire and requalify in lower grades, the new qualifications do not become effective until the dates on which the previous qualifications expire. Such qualifications remain in effect only for one year from the dates on which made.

517

Qualification effective.—Any qualification attained after the previous qualification has expired takes effect on the date on which it is made.

518

Men qualified in Army.—When men claiming regular classification as expert riflemen, sharpshooters, or marksmen reenlist from the Army within three months from the date of their discharge, the fact will be reported to the Major General Commandant by their commanding officer, and such data will be furnished regarding their previous organization and the date of qualification as will enable the claim to be verified. The right of such men to continue their classification will be announced in Marksmanship Qualification Orders, but upon the production of other properly authenticated evidence, pending receipt of such orders, credit for the extra compensation from the date of reenlistment may be made upon the first succeeding pay rolls.

519

To fire for qualification only once a year.—No man may fire the record course with the same type arm for qualification more than once during any target year; this also applies to reenlisted men who have fired the course during the target year in which they were discharged.

520

Junior officers to fire.—All officers of the line of the Marine Corps, of and below the grade of major, will be required by commanding officers to fire the Army course once per year and the Navy course as prescribed in Firing Regulations for Small Arms, United States Navy.

521

Qualification, officer, duration.—The qualification of an officer remains in effect for the same period of time as that of an enlisted man.

522

(1) *Report, entries in service-record book.*—As soon as a number of officers and men have completed firing the record practice of the Army course, their qualification, if any, the total score, the date of completion of firing, and the range upon which firing is held, will be reported to the Major General Commandant, and the same facts will be entered in the service-record book of each man concerned, and such entries will be signed by the custodian of the service-record book or the officer in charge of the range practice.

(2) *Partial firing.*—If a marine has not completed the firing begun on the regular course such partial firing will not be entered in his service-record book.

523

(1) *Muster roll and service-record book entries.*—The following shall be noted on the first muster roll after the fact, a like entry made in service-record book, and the fact will be noted on the discharge certificate of the man concerned:

(*a*) Membership on Marine Corps rifle (pistol) team in national match, either as captain, coach, spotter, principal, or alternate, thus: "Participated as (captain, coach, spotter, principal, or alternate) on Marine Corps rifle (pistol) team in national team match at ——— (place), ——— (date)." "Awarded national team match medal for membership on (first, second, third, etc.) team" (if such is the case).

(*b*) Winning a prize in the national rifle (pistol) individual match, thus: "Awarded (first, second, third, etc.) prize (gold, silver, or bronze) medal, national rifle (pistol) individual match at ——— (place), ——— (date)."

(*c*) Winning a prize in the Marine Corps match, an individual match, subscribed for by the officers of the Marine Corps, thus: "Awarded (number) prize, Marine Corps match (individual) at ——— (place), ——— (date)."

(*d*) Membership on Marine Corps teams corresponding to the departmental teams authorized in Rifle Marksmanship, thus: "Awarded (gold, silver or bronze) medal for (first, etc., to twelfth) place on Marine Corps team, ——— (date). Competition held at ——— (place)."

(*e*) "Awarded distinguished marksman's badge, 19——, by reason of holding following medals ———."

(*f*) Classification of expert rifleman, sharpshooter, or marksman under Army course.

(*g*) The data required to be entered in the service-record book under Firing Regulations for Small Arms, United States Navy.

(2) The captain of the Marine Corps rifle team is charged with the duty of furnishing the information under heads (*a*), (*b*), and (*c*) to the officer on whose muster rolls the names of the respective officers or men appear and to the Major General Commandant.

(3) Information for (*d*) and (*e*) for muster rolls to be obtained from published orders.

524

Small-arms firing, Navy.—For practice under the Firing Regulations for Small Arms, United States Navy, records will be kept and reports submitted as required therein.

Prizes.

525

When men entitled to prizes are transferred, or when their pay accounts are closed for discharge, before the prizes are paid, or credited on the pay rolls, the fact will be noted in the service-record book, and the prizes will be credited at their new station or included in the amount paid on discharge. In other cases the notation of the award of prizes will not be made in the service-record book.

526

The award or prizes will not be noted on the muster roll.

527

In addition to the prizes awarded from public funds there may be appropriated from the exchange, post, or company fund in the regular manner other prizes for marksmanship competitions, either rifle, revolver, gallery, machine gun, field pieces, automatic guns, or marksmanship tests of any kind. Officers are not debarred from these competitions and may be granted prizes therein. In like manner matches may be arranged and prizes paid from funds received by donation, voluntary entrance fees, or from any proper source.

528

(1) *Officers to be present.*—When record practice under Rifle Marksmanship is held an officer will be present at the targets, if practicable, to supervise the marking. When no officer is available for this duty the officer in charge of the

firing will take such precautions as may be in his power to insure fair and accurate marking.

(2) The presence of an officer at the firing point during record firing will never be dispensed with, and the officer should preferably be the company commander or other officer responsible for the instruction of the men.

529

The practice season for the Marine Corps will be from January 1 to December 31, inclusive, of each year.

Rifle and Pistol Competitions.

530

(1) *Kind of competitions.*—The following rifle and pistol competitions will be held each year at such times as may be designated by the Major General Commandant: Eastern Division rifle and pistol competitions at Quantico, Va.; Southeastern Division rifle and pistol competitions at Parris Island, S. C.; Western Division rifle and pistol competitions at Mare Island, Calif.; West Indies Division rifle and pistol competitions at Guantanamo Bay, Cuba; Marine Corps rifle and pistol competitions at Quantico, Va.; Elliott Trophy match at Quantico, Va.; San Diego Trophy match at Mare Island, Calif.; Lauchheimer Trophy match at Quantico, Va.

(2) *Eligibility.*—All officers and enlisted men, except distinguished marksmen and distinguished pistol shots, are eligible to participate in division and Marine Corps rifle and pistol competitions. Commanding officers, in making selections, should give due regard to steadiness, good soldierly habits, and conduct, as well as to excellence in marksmanship, and all men whose past performances or range records for the current year indicate their proficiency should be given an opportunity to qualify as competitors.

(3) *Transfer of enlisted competitors.*—Upon the receipt of authorization from the Major General Commandant to send men to compete in rifle and pistol competitions, commanding officers will select such men as they desire, within the prescribed quota, and transfer them to the place of competition without further reference to Headquarters. Immediately following their transfer, report thereof should be made, showing the names of the men thus transferred.

(4) *Transfer of officers.*—Commanding officers desiring to send officers of their commands to the competitions will telegraph their names to the Major General Commandant and request orders. Officers sent to competitions will be in addition to the quotas assigned the various posts. The detail of officers to attend this competition is desirable and should be made wherever practicable.

(5) *Distinguished rifle and pistol shots,* while not eligible to shoot for medals in division and Marine Corps rifle and pistol competitions, should be encouraged to shoot for places in order that they may be considered in connection with assignment to the team squad and the award of the Lauchheimer Trophy. Scores made by distinguished rifle and pistol shots will be included in the official report of competitions. Men in the distinguished class, when sent to competitions, will be in addition to the quotas assigned the various posts.

(6) *Officers in charge.*—The officer in charge of each competition will be designated by the Major General Commandant, and the competitions will be conducted in accordance with the general provisions outlined in Chapter VIII of Rifle Marksmanship, and Chapter IV, Part IV, of Pistol Marksmanship, and such special instructions as may be issued from Headquarters, Marine Corps.

(7) *Preliminary practice.*—In order that no officer or enlisted man may be handicapped by lack of practice, competitors will be sent to the place of com-

petition at least one week in advance of the date set for the beginning of the competition proper for preliminary shooting.

(8) *Arms.*—The United States magazine rifle, model of 1903, fitted with the No. 10 front and rear sight, and the Colt automatic pistol, caliber .45, will be used in all rifle and pistol competitions. Alterations in these weapons as issued or the use of special sights will not be permitted. In preliminary rifle firing and the competition proper, either the open or peep sight may be used at all ranges.

(9) *Ammunition.*—Standard service rifle and pistol ammunition, of the same lot if possible, will be issued by the officer in charge of the competition, and will be used by all competitors.

(10) *Uniform.*—In all rifle and pistol competitions uniform clothing and field belts will be worn. The use of leggings will be optional.

(11) *Rapid fire.*—In rapid fire all loading will be from the belt.

(12) *Reports.*—As soon as possible after the completion of competitions officers in charge will prepare reports and submit them to the Major General Commandant, showing the names of officers and enlisted men entered, medals awarded and to whom, dates of enlistment, together with such remarks and recommendations as may be deemed pertinent. These reports will also show the scores of each competitor at each range fired, and the total scores at all ranges. The scores of distinguished marksmen and distinguished pistol shots will be shown at the bottom of the reports.

(13) *The presentation of medals* will be made, when practicable, at the close of competitions, and will be conducted with the ceremony and formality warranted by the importance of the occasion.

(14) *Method of procedure.*—The officers in charge of the various competitions will prescribe rules for marking and scoring, settling protests, etc., and the general method of procedure of the competitions, which will be as nearly as possible the same as those prescribed for competitions in Rifle Marksmanship and Pistol Marksmanship.

(15) *Assignment to relays and targets.*—In division and Marine Corps rifle and pistol competitions competitors will be reassigned to relays and targets after each stage is fired, in the order of their standing in the preceding stage.

(16) *Time.*—The endurance of the competitors is a part of the test; therefore the entire program should be fired in one day, if possible. In the event that it is not possible to complete the competition proper in one day, the day's firing will cease at the completion of a stage, in order that each stage may be completed for the competitors under as nearly the same conditions as possible.

DIVISION RIFLE COMPETITIONS.

(17) *Entry list.*—The number of entries, commissioned and enlisted, in each division rifle competition shall be determined each year by the Major General Commandant. The entries will ,in all cases, be those men representing the quotas assigned the various posts by the Major General Commandant. If, for any reason, it is found impracticable to assemble the prescribed quota of enlisted men, commanding officers will inform the Major General Commandant by wire, and the competition will be delayed pending the receipt of instructions.

(18) *The preliminary practice* course will be that designated for competition in rifle marksmanship, and will be fired at least one time by each entrant prior to the firing of the competition proper.

(19) *The competition proper* will consist of the course as outlined in paragraph 209, page 132, Rifle Marksmanship, and the course will be fired through twice.

(20) *Stages.*—The order of sequence in firing will be determined by the officer in charge of the competitions.

(21) *Medals.*—Gold, silver, and bronze medals, in numbers to be determined by the Major General Commandant, will be awarded the enlisted competitors making the highest scores in each division rifle competition. The 15 winners

shall comprise the division rifle team. Commissioned competitors (except distinguished marksmen) making a score equal to that of any enlisted competitor winning a medal shall receive a medal similar in all respects to that awarded the enlisted man, and be carried as extra numbers on the division rifle team.

MARINE CORPS RIFLE COMPETITION.

(22) *Eligibility.*—Only those officers and enlisted men who were awarded medals as a result of division rifle competitions fired prior to the Marine Corps rifle competition and in the same calendar year, distinguished marksmen, and such officers and men as may be specially designated by the Major General Commandant shall be eligible to enter this competition.

(23) *Transfers.*—Upon the completion of each division competition the name of the commissioned competitors winning places in these competitions will be telegraphed to Headquarters, Marine Corps, with the request for further instructions. Upon the receipt of this telegram orders will be issued to transfer such of these officers as may be considered available to Quantico, Va., to participate in the Marine Corps competition. Commanding officers under whose jurisdiction each division competition is held are authorized to direct the post quartermasters to furnish the necessary transportation to all enlisted competitors winning medal places in the division competition, together with such commissioned competitors as may be directed by Headquarters, Marine Corps, and to order the travel necessary.

(24) *Prelminary practice* for the Marine Corps rifle competition will be the same as that prescribed for division rifle competitions, and will be fired at least one time by each competitor prior to the competition proper.

(25) *Competition proper.*—The course prescribed for division rifle competitions will be fired.

(26) *Stages.*—The Marine Corps rifle competition will be fired by stages in the same manner as division rifle competitions.

(27) *Medals.*—Gold and silver medals, in numbers to be determined by the Major General Commandant, will be awarded to the enlisted competitors making the highest scores in this competition. These 12 men will be assigned to the Marine Corps team squad for further training, with a view to their participation in the national matches if found qualified. Commissioned competitors (except distinguished marksmen) making a score equal to that of any enlisted man winning a medal will receive a medal similar in all respects to that awarded the enlisted man, and, where possible, will be assigned to the Marine Corps team squad.

ELLIOTT TROPHY MATCH.

(28) *Eligibility.*—Each post in the Marine Corps with an authorized strength of 50 men or more shall enter a team of 8 firing members in this competition. Distinguished marksmen and officers may compete in this match on even terms with other participants, no restriction being imposed upon them because of the fact that they are distinguished in the use of the rifle. In order that posts not having a rifle range situated thereon may be given the same opportunity to win this match as posts having ranges, no man who is or has been a member of an organized rifle range detachment for the four months preceding the match shall be eligible to compete. Team captains shall certify to the eligibility of their teams to the officer in charge before the match.

(29) *Time.*—The Elliott Trophy match will be held each year on the rifle range at Quantico, Va., immediately following the Marine Corps competitions. Participants in division and the Marine Corps rifle and pistol competitions shall be eligible to fire in this match as members of the teams representing their respective posts. The membership of each post team, however, is a matter for each comanding officer to determine.

(30) *The competition proper* shall consist of the complete firing of the Army qualification course, plus 2, sighting shots and 10 shots for record at 1,000 yards.

(31) *Award.*—The team attaining the highest aggregate score in this competition will be awarded the Elliott Trophy, which shall be engraved with the name of the post represented by the winning team and be held by the commanding officer of such post until the next competition is held.

SAN DIEGO TROPHY MATCH.

(32) *Eligibility.*—Each Marine Corps post on the west coast of the United States with an enlisted strength of 50 men or more, and the Pacific Advanced Base Force and Marine Barracks, Naval Station, Pearl Harbor, Hawaii, shall enter a team of 8 firing members in the competition. Distinguished marksmen and officers may compete in this match on even terms with the other participants, no restriction being imposed upon them because of the fact that they are distinguished in the use of the rifle. In order that posts not having a rifle range situated thereon may be given the same opportunity to win this match as posts having ranges, no man who is or has been a member of an organized rifle-range detachment for the four months preceding the match shall be eligible to compete. Team captains shall certify to the eligibility of their teams to the officer in charge before the match.

(33) *Time.*—The San Diego Trophy match will be held each year on the rifle range at Mare Island, Calif., immediately following the Western Division competition. Participants in the division competitions, both rifle and pistol, shall be eligible to fire in the match as members of the teams representing their respective posts. The membership of each post team, however, is a matter for each commanding officer to determine.

(34) *The competition proper* shall consist of the complete firing of the Army qualification course, plus 2 sighting shots and 10 shots for record at 1,000 yards.

(35) *Award.*—The team attaining the highest aggregate score in this competition will be awarded the San Diego Trophy, which shall be engraved with the name of the post represented by the winning team, and be held by the commanding officer of such post until the next competition is held.

DIVISION PISTOL COMPETITIONS.

(36) *Entry list.*—The number of entries, commissioned and enlisted, in each division pistol competition shall be determined each year by the Major General Commandant. The entries will, in all cases, be those men representing the quotas assigned the various posts by the Major General Commandant. When commanding officers so desire the entire quota for the pistol competition, or any part of it, may be comprised of men selected to represent their posts in the division rifle competition. If, for any reason, it is found impracticable to assemble the prescribed quota of enlisted men, commanding officers will inform the Major General Commandant by wire and the competition will be delayed pending the receipt of instructions.

(37) *The preliminary course* as outlined in Chapter IV, Part IV, of Pistol Marksmanship will be fired at least one time by each entrant prior to the firing of the competition proper.

(38) *The competition proper* will consist of the firing by each competitor of the course prescribed for preliminary practice, except that the course will be fired through twice.

(39) *Medals.*—Gold, silver, and bronze medals, in numbers to be determined by the Major General Commandant, will be awarded the enlisted competitors making the highest scores in each division pistol competition. The men winning medals shall comprise the division pistol team. Commissioned competitors making a score equal to that of any enlisted competitor winning a medal shall receive a medal similar in all respects to that awarded the enlisted man, and be carried as extra numbers on the division pistol team.

MARINE CORPS PISTOL COMPETITION.

(40) *Eligibility.*—Only those officers and enlisted men who were awarded medals as a result of division pistol competitions fired prior to the Marine Corps pistol competition and in the same calendar year, distinguished pistol shots, and such officers and men as may be specially designated by the Major General Commandant, will be eligible to enter this competition.

(41) *Transfers.*—The transfers of medal winners in the division pistol competition will be effected in the same manner as prescribed above for medal winners in the division rifle competition.

(42) *Preliminary practice* for Marine Corps pistol competitions will be the same as that prescribed for division pistol competitions, and will be fired at least one time by each competitor, prior to the competition proper.

(43) *Competition proper.*—The course prescribed for division pistol competitions will be fired.

(44) *Medals.*—Gold and silver medals, in numbers to be determined by the Major General Commandant, will be awarded the enlisted competitors making the highest scores in the Marine Corps pistol competition. These men shall be assigned to the Marnie Corps team squad for further training, with a view to their participation in the National matches, if found qualified. Commissioned competitors (except distinguished pistol shots) making a score equal to that of any enlisted competitor receiving a medal shall be awarded a medal similar in all respects to that awarded the enlisted man, and, where possible, will be assigned to the Marine Corps team squad.

LAUCHHEIMER TROPHY MATCH.

(45) *Conditions of award.*—The Lauchheimer Trophy will be awarded annually to the officer or enlisted man who makes the highest final score, computed in accordance with the formula given below, with both rifle and pistol during the Marine Corps competitions. The winner will receive a letter of commendation from the Major General Commandant and will be considered the individual shooting champion of the Marine Corps for the year.

(46) *Men in distinguished class.*—Distinguished marksmen and distinguished pistol shots are eligible to receive this trophy, and the scores made by them while shooting in Marine Corps competitions will be considered in connection with its award.

(47) *Method of determining the winners.*—Immediately after the completion of the Marine Corps rifle and pistol competitions, referred to above, the total rifle and pistol scores of each competitor, both commissioned and enlisted, will be computed as indicated below:

$$\left.\begin{array}{l}\text{Total rifle score times 67.8}\\ \text{Total pistol score times 32.2}\end{array}\right\}\text{Divide by 100,}$$

and the competitor attaining the highest final score will be declared the winner, and will receive a gold medal and a letter of commendation from the Major General Commandant. The name of the winner will be engraved upon the trophy, which is emblematic of the rifle and pistol championship of the Marine Corps, and which, under the conditions of award, must be kept in the office of the Major General Commandant at Headquarters, U. S. Marine Corps.

531

(1) *Ammunition.*—There shall be kept on hand at all Marine Corps posts, and at all powder magazines and radio stations where marines are stationed, not less than 100 rounds caliber .30 ball ammunition per man, of authorized strength. This does not contemplate the reduction of the ammunition reserve at any post where a larger amount has been or may be specifically designated.

Commanding officers should maintain a larger reserve where military considerations require.

(2) *Oldest ammunition to be used in preliminary practice.*—At all posts in the Marine Corps which have rifle ranges attached or in the immediate vicinity thereof, the commanding officer will cause the preliminary target practice of all marines firing on the range under his jurisdiction to be held with the oldest lot of ammunition on hand in order to prevent the deterioration of the ammunition supply.

532

(1) *Marine Corps distinguished marksman.*—Any officer or enlisted man of the Marine Corps who since 1906 has won a medal in a division or Marine Corps competition, and, including such medal, has won any three of the following authorized medals, shall be classed as a distinguished marksman and awarded a medal as such:

Division competition, Marine Corps.
Marine Corps competition.
National individual match.
National team match, as a shooting member of the highest team representing the Marine Corps.
Expert team rifleman, U. S. Navy.

(2) In addition to the above medals any officer or enlisted man of the Marine Corps who while serving in the Army, or while detached for service with the Army, has been awarded a badge for representing any branch of the Army on a team at a National match as a shooting member, or has won any of the badges authorized in Rifle Marksmanship, as a credit towards a distinguished marksman, shall include such badge among those which will count in determining eligibility for classification as a distinguished marksman, Marine Corps.

(3) Officers and enlisted men of the Marine Corps who have won three of the above medals will make application to the Major General Commandant for classification as a distinguished marksman, giving full data concerning their eligibility for such classification, in order that a medal may be awarded.

CHAPTER 6.

PROMOTION AND REDUCTION.

Section 1.—OFFICERS.

EXAMINATION FOR PROMOTION.

601

(1) *Procedure.*—Boards for the examination of officers of the Marine Corps for promotion shall be governed by the procedure laid down in Naval Courts and Boards, latest edition.

(2) *Publications used.*—The board will base its questions on the latest edition of the following publications, except as noted in paragraph 725, Naval Courts and Boards, 1917:

Navy Regulations.
Navy Department General Orders.
Marine Corps Orders and Instructions.
System of Accountability.
Instructions Governing Transportation of Troops and Supplies for U. S. Marine Corps.
Manual of the Paymaster's Department, U. S. Marine Corps.
Military Topography—Sherrill.
Army Regulations.
Naval Ordnance, U. S. Naval Academy—1921.
Naval Courts and Boards.
Elements of International Law—Davis.
Field Service Regulations.
Deck and Boat Book, U. S. Navy.
Infantry Drill Regulations (Provisional), U. S. Army, 1919.
Bayonet Training Manual, U. S. Army.
Landing Force Manual, U. S. Navy, 1920.
Rifle Marksmanship, U. S. Army.
Pistol Marksmanship, U. S. Army.
Musketry Bulletins, A. E. F. (1919).
Description and Rules for the Use of Musketry Rule Model, 1917.
Engineer Field Manual, 1917.
Engineer Field Notes, A. E. F.
Pamphlet on Equitation, General Service Schools.
Horses, Bridles, and Saddles—Carter.
Army Regulations.
Rules of Land Warfare.

(3) *The scope of professional examinations* shall be as follows:

(4) *For officers of the Quartermaster's Department:*
(*a*) Administration of Quartermaster's Department.
(*b*) General efficiency.

(5) *For officers of the Paymaster's Department:*
(*a*) Administration of the Paymaster's Department.
(*b*) General efficiency.

(6) *For officers of the line:*
(A) *For promotion from second lieutenant to first lieutenant*—
(*a*) Administration—
Navy Regulations, Chapters 2, 3, 4, 5, 16, 17, 36, 39, 44, 45, 46, 47, and 52.
Navy Department General Orders in force applicable to the Marine Corps.
Marine Corps Orders.
System of Accountability.
Manual of Paymaster's Department.
Instructions governing the transportation of troops and supplies.
The preparation of all rolls, returns, and papers connected with a detachment and company.
The examination shall be sufficiently extended to determine whether the candidate is entirely familiar with the subject generally, as well as his own duties, arising under the regulations and orders specified, both on shore and on board ship.
(*b*) Drill Regulations—
Landing Force Manual, U. S. Navy, 1920, as follows:
Part I, Chapter I, The Landing Force.
Part I, Chapter II, Interior Guard Duty.
Part I, Chapter IX, First Aid and Military Hygiene.
Infantry Drill Regulations (Provisional), 1919, U. S. Army, as follows—
Definitions; introduction; orders, commands, and signals.
School of the Platoon, close and extended order.
School of the Company, close order.
Bayonet Training Manual, U. S. Army.
Practical: The candidate will be required to demonstrate his ability to handle a company or platoon in close and extended order.
(*c*) Marksmanship:
Rifle Marksmanship.
Pistol Marksmanship, omitting Part III.
(*d*) Signals, Practical: To be sufficiently extended to demonstrate ability of candidate to send and receive messages by (1) Army and Navy Code (wigwag), and (2) Two Arm Semaphore System.
(*e*) Musketry:
Musketry Bulletins, American Expeditionary Forces, 1919; Descriptions and Rules for the use of the Musketry Rule, model of 1917; Musketry, Practical: Such individual demonstration without troops as the board may require to show that the candidate has a practical knowledge of musketry.

(6) *For officers of the line:*—Continued.

(A) *For promotion from second lieutenant to first lieutenant*—Continued.

(*f*) Military Field Engineering:

Field Engineer's Manual, U. S Army, Part II, pars. 1, 2, and 46–66, inclusive, omitting tables. Part V, pars, 24–36, 49–64, 119–131. Engineer's Field Notes, A. E. F.: Nos. 2, 3, 12, 14, 19, 20, 23, 24, 26, 27, 28, 29, 31, 32, 33, 34, 35, 38, and 42.

(*g*) Naval and Military Law:

Naval Courts and Boards, latest edition; the examination shall be such as to determine whether the candidate is entirely familiar with the duties of a recorder of courts and boards, the most common rules of evidence and the powers and limitations of officers in connection with the administration of justice in the Marine Corps.

(*h*) Tactics:

Infantry Drill Regulations (Provisional) 1919, U. S. Army: Combat and Offensive Combat: Field Service Regulations, U. S. Army, Part II, Articles I (Information); II (Security); III (Orders); and IV (Marches), omitting Convoys by Water; Engineer Field Notes, American Expeditionary Forces; E. F. N. No. 13: Landing Force Manual, U. S. Navy, 1920; Part I, Chapter VII (Minor Warfare). Practical: A suitable number of troops being assigned and the desired terrain being available, the candidate shall be required to work out in the vicinity of the post a practical problem when in command of a platoon in offensive combat. If this is not practicable the board shall so state, with reasons therefor, in its record of proceedings and shall require the candidate to solve a map problem involving offensive combat where the force at his disposal shall not exceed a company, the situations given and the solutions thereto to be written and appended to the report.

(*i*) Naval Ordnance and Gunnery. For all second lieutenants: Naval Ordnance, U. S. Naval Academy, 1921 edition, Arts. 27–54, 120, 142 175, 356–373, 427–500, 506–559 (omitting Turret Gun Sights), 560–579 (omitting Turret Firing Circuits and Gas-Expelling Devices for Turret Guns), 617–722, 817–819.

(*j*) Boats: Deck and Boat Book, U. S. Navy: Boats in General; Boat Gear and Equipment. Life Boats; Drills and Exercises; Detailed Notes on Boat Duty; Boat Salutes, etc.; Rules of the Road.

(*k*) Military Topography: Military Topography, Sherrill, Parts I and III. Practical: A position or road sketch. The Board will select the ground, assign the limits to be covered, and establish the time limit. Upon completion of the field work the officer will at once report to the officer in charge of the examination room with the result of his work, and complete the necessary maps, reports, etc.

(*l*) General efficiency.

(6) *For officers of the line:*—Continued.

(B) *For promotion from first lieutenant to captain*—

(*a*) Administration: Same as for promotion to first lieutenant; in addition, the examination shall be sufficiently extended to determine whether the candiate is familiar with and has the necessary knowledge to perform the duties of a post exchange officer, a post quartermaster, a regimental quartermaster, and a regimental paymaster.

(*b*) Drill Regulations:

Landing Force Manual, U. S. Navy, 1920, Part I, Chapter I. Infantry Drill Regulations (Provisional), 1919, U. S. Army.

School of the Company.

Practical: The candidate will be required to demonstrate his ability to handle a company in close order.

(*c*) Naval and Military Law: Naval Courts and Boards, latest edition. The examination shall be such as to determine whether the candidate is entirely familiar with the whole subject matter of the text in so far as it applies to the Marine Corps.

(*d*) Rules of Land Warfare.

(*e*) Military Field Engineering: Field Engineer's Manual, U. S. Army: Part II, pars. 1, 2, and 46–66, inclusive, omitting tables. Part V, pars. 24–36, 49–64, 119–131, Engineer's Field Notes, A. E. F.; Nos. 2, 3, 12, 14, 19, 20, 23, 24, 26, 28, 31, 32, 33, 34, 35, 38, and 42.

(*f*) Tactics: Infantry Drill Regulations (Provisional), 1919, U. S. Army; Combat and Offensive Combat; Field Service Regulations, U. S. Army: Parts II, except Convoys by water; Engineer Field Notes, American Expeditionary Forces; E. F. N. No. 13; Landing Force Manual, U. S. Navy, 1920, Part I, Chapter VII (Minor Warfare). Practical: The candidate will be given a practical problem in some phase of combat or security to solve on varied terrain in the vicinity of the post when in command of a company of infantry. If this is not practicable the board shall so state, with reasons therefor, in its record of proceedings and shall require the candidate to solve a map problem involving offensive combat or security where the force at his disposal shall not exceed a battalion of infantry, the situations given and the solutions thereto to be written and appended to the report.

(*g*) General efficiency.

(C) *For promotion from captain to major*—

(*a*) International Law—Davis. Omit chapters V, VI, and VIII, and appendices; Field Service Regulations, appendix 6.

(*b*) Tactics: Infantry Drill Regulations (Provisional), 1919, U. S. Army; Combat and Offensive Combat; Field Service Regulations, U. S. Army, omitting appendices; Engineer Field Notes, American Expeditionary Forces, E. F. N. No. 13; Landing Force Manual, U. S. Navy, 1920, Part I, Chapter VII (Minor Warfare). Practical: The candidate will be required to solve a map problem involving combat or security where the force at his dis-

(6) *For officers of the line:*—Continued.

(C) *For promotion from captain to major*—Continued.

(*b*) Tactics:—Continued.

posal shall not be greater than a regiment of infantry plus a proportionate addition of other arms. The problem to be such as to require an estimate of the situation, preparation of the necessary orders, and a tracing showing disposition of troops at a certain phase in the situation as determined by the board and announced to the candidate. The problem and the written solution shall be appended to the report.

(*c*) Administration: Navy Regulations; Navy Department General Orders; Marine Corps Manual; Marine Corps Orders; System of Accountability; Manual of the Paymaster's Department. The examination shall be sufficiently extended to determine whether the candidate is familiar with and has sufficient knowledge to perform the duties of a quartermaster of an expeditionary brigade.

(*d*) Equitation—

General Service Schools pamphlet.

Hippology: Horses, Saddles, and Bridles—Carter.

(*e*) Rules for Land Warfare.

(*f*) General efficiency.

(D) *For promotion to lieutenant colonel and colonel*—

(*a*) Tactics. Practical: The candidate to be required to solve a map problem involving the higher functions of staff duties and command. The problem and the written solution shall be appended to the report.

(*b*) General efficiency.

(7) *Election to be examined under Marine Corps Order No. 25 (Series 1920).*—Prior to July 1, 1922, officers may elect to be examined under the provisions of Marine Corps Order No. 25 (Series 1920) as originally issued.

EXEMPTIONS.

602

(1) *Service schools.*—When a candidate for promotion has attended and completed a regular course at any Army, Navy, or Marine Corps school or college for commissioned officers, except post schools, his certificate of graduation therefrom will be accepted in lieu of examination in the subjects in which graduated, upon examination for promotion next after the date of graduation. In addition, certificates of graduation already issued, or hereafter issued to officers now attending the courses, will be accepted in lieu of such examination, in accordance with section 729, Naval Courts and Boards, as originally published.

(2) *Prior examination in special subjects.*—The marine examining board, Headquarters, U. S. Marine Corps, will examine an officer, upon application, in any subject or subjects, except general efficiency, required for professional examination for promotion to the next higher grade or rank, and if the officer passes, will issue a certificate of qualification, which will be accepted upon his next examination for promotion in lieu of professional examination in such subject or subjects. The board will not, under the provisions of this paragraph, again examine an officer in any subject in which he fails to pass prior to his next promotion, but such failure will not deprive the officer of the right to the usual examination for promotion, as provided by law. When an officer does not appear before the board, written examination will be conducted before a

supervisory officer, upon questions prepared and marked by the board, and practical exercises designated by the board will be conducted and marked by a subsidiary examining board of one or more officers, appointed by the Major General Commandant, or by his authority.

(3) *Diplomas from schools and colleges.*—The marine examining board, Headquarters, U. S. Marine Corps, will also, upon application, consider a diploma or certificate received by an officer from any school, college, or university for proficiency in any professional subject or subjects required for his examination for promotion to the next higher grade or rank, and if satisfied that the diploma or certificate is evidence of qualification in such subject or subjects equivalent to that of passing the usual examination, will issue a certificate of qualification, which will be accepted upon his next examination for promotion in lieu of professional examination in such subject or subjects.

Probationary Officers.

603

(1) *The competitive examination* provided by the act of August 29, 1916, for officers at the end of their probationary period, will be conducted by the marine examining board, Headquarters, U. S. Marine Corps.

(2) *The examination will be written,* and will include all subjects required for written examination for promotion to first lieutenant. The questions will be prepared by the examining board, and will be answered by the competitors in the presence of supervising officers, at the place of duty.

(3) *The complete records of candidates,* with all papers pertaining thereto, will be submitted to the board for consideration in connection with marks to be assigned under subject "General efficiency."

(4) *The weights* assigned to each subject in determining the final mark on which relative standing is based, will be as follows:

Subject	Weight	Subject	Weight
Administration	3	Naval and military law	3
Drill Regulations	3	Naval ordnance and gunnery	2
Signals	1	Tactics	5
Musketry	2	General efficiency	10
Military field engineering	1	Boats	1
Military topography	3	Marksmanship	1

(5) *The board will issue a certificate* to each probationary officer examined, stating the mark attained in each subject, which will be accepted upon the next examination for promotion in lieu of professional examination in subjects, except general efficiency, in which the mark is 2.5 or over.

Examining Boards.

604

(1) *Supervisory examining boards* will be appointed by the Major General Commandant, or by his authority, and *supervising officers for examinations* will be designated by the commanding officer upon request by a marine examining board. The following rules will govern:

(2) The candidate will be furnished only such number of questions or be required to conduct such practical exercises as he may be able to answer or complete before a recess or adjournment is taken. Examinations shall continue from day to day, Sundays and holidays excepted, until completed.

(3) The candidate will be required to submit, at the completion of the examination, a certificate in the form prescribed by Naval Courts and Boards for examination of marine officers that he has received no unauthorized assistance.

(4) All papers will be forwarded by the supervisory board or supervising officer, direct to the marine examining board having cognizance, as soon after the completion of the examination as may be practicable.

(5) Supervisory boards and supervising officers will be further governed by the instructions of the marine examining board having cognizance.

(6) *Examination questions, preparation of.*—The Marine examining board, Headquarters, Washington, D. C., will, in order to make the examination of officers of the several grades and ranks uniform, prepare examination questions as guides for other marine examining boards which may be ordered.

Graduation.

605

(1) *Graduates from any class of the Marine Corps schools* shall have the fact of graduation entered on their official record.

(2) *Navy Register.*—The names of officer graduates will be borne on the Navy Register with the note "Graduate Marine Corps schools," followed by the course from which graduated, as "Field officers' course."

Section 2.—ENLISTED MEN.

Appointment to the Naval Academy.

606

(1) *Eligibility.*—Enlisted men of the Marine Corps are eligible for appointment to the Naval Academy under the following instructions in the Bureau of Navigation Manual which are published for the information of all concerned. Applications for leave or for detail to one of the service schools referred to in articles D–5000 and D–5010 will be sent through official channels to the Major General Commandant.

(2) *Instructions.*—" D–5000. Enlisted men who receive congressional appointments as principals or alternates for appointment to the Naval Academy or to the Military Academy, upon their own applications, submitted through regular channels, may be granted leave in order to return to their homes and make special preparation at private schools for entrance examinations. Such requests will be approved only by the bureau. Leave granted under this authority will expire 10 days after the completion of the examinations.

" D–5010. Men who do not desire to take leave as provided in article D–5000 may, upon application through official channels to the bureau, be detailed to one of the service schools referred to in article D–5015.

" D–5011. The law authorizes the appointment to the Naval Academy each year of 100 midshipmen, to be selected as a result of the competitive examination given enlisted men of the regular Navy and Marine Corps, members of the Naval Reserve Force, and Marine Corps Reserve on active duty. Candidates must be not over 20 years of age on April 1 and have been in the service at least one year by August 15 of the year of entrance. The mental and physical requirements for these candidates are the same as for other candidates for appointment as midshipmen. The examinations will be held in April of each year at the naval training stations, San Francisco, Calif., and Newport, R. I., on dates designated by the bureau.

" D–5012. Men who desire to take the examination must have the following qualifications:

" (*a*) Must have enlisted in the regular Navy or Marine Corps or in the reserve of either branch of the service on or before August 15 of the year preceding the examination.

and must have passed successfully at least two half-years of algebra and one half year of Geometry.

"(b) Must be not less than 16 nor more than 20 years of age on April 1 of the year in which they take the examination.

"(c) Must be citizens of the United States.

"(d) Must be able to pass a rigid physical examination.

"(e) Must have had two years in high school or equivalent education.

"(f) Must be of officer caliber.

"D-5013. Commanding officers of all ships and stations will ascertain the names of all enlisted men in their commands who are eligible and who desire to compete in the examinations. They will have each candidate given a preliminary physical and mental examination, and none who is manifestly unfitted to be a competitor should be considered. The names of those selected will be submitted to the Bureau of Navigation, in duplicate, on the following form, arranged in order of merit:

Name in full (surname first).	Rate.	Date enlisted.	Date and place of birth.
....	....	....	....
....	....	....	....
....	....	....	....
....	....	....	....
....	....	....	....

"The necessary steps should be taken by all those charged with the responsibility of selecting candidates to see that no eligible candidate is overlooked.

"D-5014. On the receipt of the above-described list, which must be submitted prior to December 1 of each year, the bureau will direct to which school the candidates will be transferred. In case there is a possibility of any list not reaching the bureau from cruising ships or from stations without the continental limits of the United States by December 1, the list of names will be telegraphed to the bureau.

"D-5015. To provide for intensive instruction of these candidates, special schools are established at the naval training stations, ~~Newport, R. I.~~, Hampton Roads and ~~San Francisco~~ San Diego, Calif. The course of instruction will begin in ~~January~~ on a date set by the bureau, and will continue until the date set for the beginning of the competitive examination. It is desired that as large a number of men as it may be possible to obtain under the law and the required mental and physical qualifications be selected. The course of instruction shall be such as to prepare each candidate for the entrance examination as described in the "Regulations Governing the Admission of Candidates into the United States Naval Academy as Midshipmen" and "Examination Papers." Copies of these pamphlets will be forwarded to the training stations concerned. Existing facilities and equipment should be utilized to the fullest possible extent. Textbooks not already available must be supplied by the students. On completion of the final competitive examinations, the completed examination papers will be forwarded by the commandants concerned direct to the Superintendent, Naval Academy, Annapolis, Md., where the papers will be marked. Students under instruction will be kept under strict military discipline and required to observe the rules and regulations of the station; but all due consideration shall be given to the purpose for which they are detailed to this three months' special duty, and there shall be as little interference as possible with the schedule of school work."

NONCOMMISSONED OFFICERS.

WARRANTS.

607

(1) *Classes.*—Warrants are divided into two general classes—regular and technical. Regular warrants are either temporary or permanent. All technical warrants are temporary.

(2) *All warrants shall be probationary* for a period of six months, and while probationary are subject to revocation for cause by appointing powers. At the expiration of the probationary period of six months probationary warrants shall, whether regular or technical, either be revoked by the appointing power or forwarded to the Major General Commandant (or the departmental commander, in the Department of the Pacific) with recommendation for confirmation. Confirmed warrants shall be revoked only by sentence of court-martial or by the Major General Commandant or departmental commander.

(3) *Technical warrants* will be issued for technical duties and will state the nature of such duties, as "technical warrant for duty as clerk," "technical warrant for duty as plumber," "technical warrant for aviation duty," etc. They will contain no notation regarding duty with a specified organization.

(4) *Revocation of probationary warrants.*—Upon the cessation of the detail for the duty for which issued, probationary warrants, whether regular or technical, will be revoked by the appointing power, and confirmed temporary warrants will be referred to the Major General Commandant or to the departmental commander for instructions, except that in case of transfer a temporary warrant will continue to the new post of duty, where, if there be no vacancy for the noncommissioned officer in the capacity for which appointed, the commanding officer will revoke the warrant, if probationary, or refer it to the Major General Commandant, or the departmental commander, if confirmed.

(5) *Transfer for discharge.*—A technical noncommissioned officer transferred to the United States for discharge will retain and be discharged in his technical rank.

(6) *Sergeants major* receive probationary regular warrants, which upon confirmation are permanent.

(7) *Quartermaster sergeants, sergeants, and corporals* receive probationary regular warrants, which upon confirmation become permanent, or probationary technical warrants, which upon confirmation are temporary.

(8) *First sergeants* receive probationary regular warrants, which upon confirmation are temporary.

(9) *Gunnery sergeants* receive probationary technical warrants, which upon confirmation are temporary.

EXAMINATIONS.

608

(1) *Examining board.*—In all cases of promotion of privates or noncommissioned officers the commanding officer of marines shall convene a board to conduct the prescribed examination, except where no examination is required.

(2) *Board for technical appointments.*—Examining boards for candidates for technical appointments shall be composed as nearly as may be practicable of field officer of the regiment or other organization, and of two officers of the company or detachment in which the appointment is to be made. The examination shall consist of such technical subjects as will demonstrate the candidates' fitness for the duties to be required of them.

(3) *Corporals.*—The examination of privates to be corporals shall consist of reading, writing, and the simple rules of arithmetic, a knowledge of the duties of a corporal, and of the school of the soldier, of the squad in close and extended order, and of the platoon, and the manual of guard duty.

(4) *Sergeants.*—The examination of corporals to be sergeants shall consist of the examination of privates to be corporals, with the addition of a knowledge of the company in extended order drill, the duties of a sergeant, and the keeping of necessary accounts, making out muster rolls and the various blanks and returns required to be rendered by a sergeant in charge of a detachment.

(5) *Quartermaster sergeants.*—Candidates for appointment to the grade of quartermaster sergeant (regular warrant) shall be examined in arithmetic, composition of official letters, spelling, and typewriting. In addition, they shall be examined in accountability and administration of the staff department (adjutant and inspector's, quartermaster's, or paymaster's) to which they are candidates for appointment for duty, and in the use and preparation of blank forms pertaining to that department.

(6) *Sergeants major* should be selected from men of considerable military experience, preferably from the list of first sergeants or gunnery sergeants, and may be appointed without examination.

(7) *First sergeants* should be selected from the list of gunnery sergeants or sergeants, and may be appointed without examination.

(8) *A reenlisted marine* who at the time of his discharge from the Marine Corps was a noncommissioned officer may be warranted in the noncommissioned rank he held at the time of his discharge by the Major General Commandant without examination.

(9) *Duties of gunnery sergeants.*—Gunnery sergeants shall not be detailed as clerks, orderlies, or chauffeurs, or for mess, commissary, post exchange, guard, or police duties, but they shall be detailed for such military duties as the command of sections or platoons, or to act as coaches in connection with preliminary target instruction and on the rifle range; as assistants in the instruction of the enlisted personnel in military drills, military topography, signals, patrolling outposts, field engineering, combat firing, estimating distances, machine guns, field artillery, rapid fire and other guns, portable searchlights, submarine mines, fire control, handling boats, seamanship, etc., and to take charge of arms, guns, and other military or technical material in the custody of organizations to which they are attached. A gunnery sergeant in the absence of the first sergeant, may be detailed as acting first sergeant.

General.

609

(1) *Complements of noncomissioned officers* for posts and organizations will be announced from time to time by the Major General Commandant and shall not be exceeded.

(2) If the organization is not up to authorized strength the number of noncommissioned officers will be kept in proportion to the number of privates and privates, first class, unless specially authorized by the Major General Commandant.

(3) If, through transfer or reenlistment, the noncommissioned strength exceeds the authorized or proportionate strength, such noncommissioned officers may be retained in excess until the number is reduced to that authorized.

(4) Promotions shall not be made to any grade at any post, in any battalion not a part of a regiment, in any regiment not brigaded, or in any brigade so long as there is an excess of the total number in that grade in the whole major organization over the total complement allowed.

610

(1) ***Authority to make probationary appointments*** of noncommissioned officers within authorized complements is conferred on the following:

(*a*) Commanding general Department of the Pacific.

(*b*) Commanding officers of advanced base or expeditionary forces among troops belonging to their headquarters.

(*c*) Brigade commanders among the troops belonging to their headquarters.

(*d*) Commanding officers of regularly organized regiments, separate battalions, and aviation squadrons.

(*e*) Marine officers commanding separate marine organizations on armed transports of the Navy.

(NOTE.—The authority given to commanding officers of various organizations in paragraphs (*b*) to (*d*) will be exercised by them when the organizations are on vessels of the Navy as well as on shore.)

(*f*) Commanding officers of vessels of the Navy having marine detachments as part of the ship's complement.

(*g*) Commanding officers of posts and separate detachments on shore within and without the United States.

(2) ***Blank warrants*** will be furnished, upon requisition, by the depot quartermaster, Philadelphia, or, as provided by article 927, by the Assistant Adjutant and Inspector, San Francisco.

611

Transfers.—Except in cases of emergency, a noncommissioned officer will not be transferred away from his post or station during his probationary period.

612

Warrants of those recommended for commissions.—When noncommissioned officers who are recommended for commissions are sent for training to a selected post they will be transferred to such post, but their absence will not be considered as creating vacancies in their respective grades and no promotions will be made to fill the vacancies of such men at the post from which transferred pending the result of the examination for commission.

613

Reductions will be effected by letter addressed to the noncommissioned officer, to be attached to his warrant, giving date and reason for reduction.

614

Desertion.—Should a noncommissioned officer or private first class be declared a deserter, his position shall be considered as vacated from the date of his unauthorized absence, and if he be subsequently returned to the service he shall be taken up as a private.

615

(1) ***Discharge.***—Commanding officers of marines, upon delivery of discharge certificates of noncommissioned officers or privates first class, shall always indorse thereon, under "Recommendations," a statement as to whether or not the man is recommended for reappointment to his former grade should he reenlist.

(2) Noncommissioned officers so reappointed will immediately assume the rank, insignia, and duties, and will be entitled to the pay of their respective grades, from the date of such reappointment.

(3) An enlisted man shall not be advanced in rank for the purpose of discharge. The advanced rank gives certain rights to increased bounty upon reenlistment, and conferring the advanced rank just prior to discharge gives the man an unearned benefit at the expense of the Government.

616

(1) *Reenlistment.*—Upon the reenlistment within three months from date of discharge of any private first class or of any noncommissioned officer who at the time of discharge was not serving under a probationary warrant, the enlisting officer shall at once reappoint him to the grade in which discharged if discharged with character "Excellent," or if discharged with character "Very good," and recommended on his discharge certificate for reappointment upon reenlistment; but if discharged with character "Very good" and his discharge certificate contains no recommendation as to reappointment, the question of his reappointment shall be referred to the Major General Commandant.

(2) If the noncommissioned officer is discharged while serving under a probationary warrant, and holds a permanent or confirmed warrant in a lower grade, the recommendation on his discharge certificate shall refer to his permanent or confirmed rank, and if reappointed upon reenlistment he shall be reappointed in such permanent or confirmed rank.

(3) Where, however, a noncommissioned officer holding a probationary warrant immediately reenlists at his place of duty and continues to perform the duty, he will be reappointed upon reenlistment to the rank in which he held a probationary warrant when discharged.

(4) Upon reenlistment, a noncommissioned officer holding a temporary warrant (regular or technical) will not be transferred to a post or organization in which there is no vacancy for a noncommissioned officer of his grade unless he elects to be reduced to his permanent grade.

(5) The reappointment of a noncommissioned officer or private, first class, shall be evidenced by a letter addressed to him by the appointing officer. This letter shall state the fact and date of reappointment and the reason therefor, such as "Upon reenlistment (date)," or "Upon relief from recruiting duty (date)," etc.

617

A copy of each appointment, reappointment, or reduction shall, as made, be forwarded to the Major General Commandant, and, if within the Department of the Pacific, an additional copy shall be forwarded to the departmental commander.

618

Abbreviations for entries in service-record books regarding kinds of warrants will be as follows:

P. wrnt	Permanent warrant.
Con. tem. reg	Confirmed temporary regular.
Prob. reg	Probationary regular.
*Prob. tech	Probationary technical.
*Con. tem. tech	Confirmed temporary technical.
S. wrnt	Ship's warrant.
R. wrnt	Recruiting warrant.

*The nature of the technical duties for which appointed should be stated as "Prob. tech. clerk," or "Con. tem. tech. mechanic."

PRIVATES, FIRST CLASS.

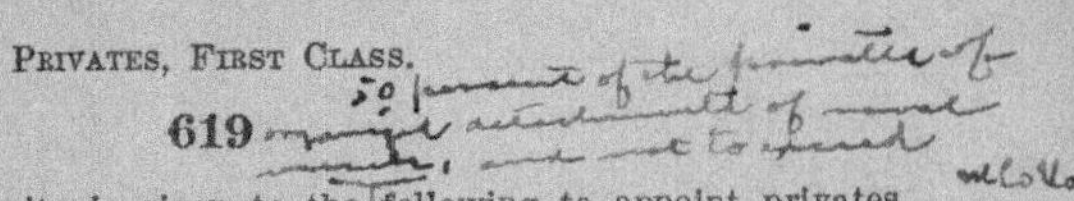

619

(1) *Appointments.*—Authority is given to the following to appoint privates first class within their commands not to exceed 20 per cent of the privates of ~~such~~ commands:

(*a*) Commanding general, Department of the Pacific.
(*b*) Commanding officers of all posts.
(*c*) Brigade commanders among the troops belonging to brigade headquarters; commanding officers of advanced base forces, regularly organized regiments, separate or detached battalions, and companies.
(*d*) Commanding officers of organized marine detachments on armed transports of the Navy.
(*e*) Commanding officers of separate expeditionary organizations, less than a regiment outside of the United States.
(*f*) Commanding officers of Marine Aviation Squadrons.
(*g*) Officers in charge of recruiting divisions, for detail on recruiting duty.
(*h*) Commanding officers of vessels of the Navy having organized marine detachments.
(*i*) Commanding officers and noncommissioned officers in charge of separate marine detachments ashore.

(2) *Reduction.*—The above appointing powers have authority to reduce first-class privates to privates, within their organizations, as they may see fit.

(3) *Transfers.*—Appointments will not be vacated by transfer from an organization. Whenever, by reason of transfer, the proportion of first-class privates exceeds ~~20 per cent~~ of the total number of privates attached to the command, the excess may be retained until vacancies occur within the organization or until the number can be reduced by transfer.

(4) *Form of appointments.*—Appointments will be made in the form of an order signed by the commanding officer.

620

Probationers.—Enlisted men who are serving in probationary periods as the result of conviction by courts-martial shall not be promoted to noncommissioned rank or appointed private first class.

CHAPTER 7.

DISCIPLINE.

PUNISHMENTS.

701

(1) *Regulations.*—Punishments for offenses committed by persons belonging to the Marine Corps shall be inflicted in accordance with the provisions of the Articles for the Government of the Navy, and the limitations prescribed by articles 24 and 25 thereof shall be strictly observed by the commanding officer of marines, and under no circumstances shall an offender be placed on guard or required to perform extra guard duty as a punishment, whether serving afloat or on shore. (When, however, marines are detached for service with the Army by order of the President, they are subject, while so detached, to the rules and articles of war prescribed for the government of the Army, sec. 1621, R. S.)

(2) *Deck and summary courts.*—For the trial of offenses which the commanding officer of marines may deem deserving of greater punishment than he is authorized to inflict under the provisions of article 24, A. G. N., but not sufficient to require trial by general court-martial, he may order a deck court or summary court-martial.

(3) *General courts-martial.*—Offenses, which, in his opinion, require the trial of the offender by a general court-martial, shall be reported by him through official channels to the officer within whose command he is serving who has authority to convene general courts-martial, or, if there be none such, to the Secretary of the Navy, through the Major General Commandant.

(4) *Accidental discharge of firearms.*—Every man who is responsible for an accidental discharge of any weapon through careless handling shall be tried by general court-martial, summary court-martial, or deck court according to the degree of negligence and the consequences of the act.

(5) *Officer under arrest.*—When a commanding officer has an officer placed under arrest for trial by general court-martial, he will notify the paymaster carrying such officer's account of his action and of the nature of the charges which are to be preferred. A copy of the notification will be furnished to the Paymaster, U. S. Marine Corps, Washington, D. C.

DESERTERS AND STRAGGLERS

702

Rewards.—An offer of reward for the apprehension and delivery of a straggler or a deserter shall state that payment of said reward will be made upon the delivery of the straggler or the deserter at any marine barracks or post or Marine Corps recruiting station, and must not be issued after the date of aphension. In case of a deserter, the offer of reward shall specify that delivery must be made on or before a certain calendar date, which shall be such as will admit of his trial by general court-martial before the statute of limitations can be invoked.

703

Desertion defined.—Absence without leave with a manifest intention not to return, shall be regarded as desertion; therefore, when the intention to desert is apparent, a reward will be offered immediately, without awaiting the expiration of 10 days.

704

(1) *The staff returns* of a deserter shall be forwarded to the Major General Commandant.

(2) *Report of apprehension, etc.*—When a deserter or straggler is apprehended or surrenders himself on board ship or at a barracks the commanding officer will immediately report the fact, through the proper channels, to the Major General Commandant.

705

Reward, where sent.—In forwarding the staff returns of any enlisted man dropped as a deserter, a notation will be made in his service record book as to whether a reward has been offered for his apprehension and delivery, and if so the amount of the reward. A copy of the reward for the apprehension and delivery of a deserter or straggler will be sent to the adjutant and inspector; to the deserter's next of kin; to the mayor (or chief of police) of the town where he resided at date of enlistment; to the chief of police of any other towns to which it is thought he may have proceeded, the selection being governed by facts obtained from his military history, declarations to comrades, or other sources of information; and to such detective agencies as may from time to time be designated by Headquarters.

706

(1) *Entries in service-record books.*—In order to have at hand and immediately available for use before general courts-martial the necessary evidence for the trial of enlisted men of the Marine Corps who have been declared deserters, and to aid courts-martial in determining whether such men are guilty of desertion or absence without leave, when the staff returns are closed entries will be made in service-record books of such of the following circumstances attending the desertion as may be known:

Disposal of effects, especially uniforms.
Secret preparations.
Procuring plain clothes.
Declarations.
Desire to quit the service.
Taking passage for a distant point.
Escaping from arrest.
Commission of an offense and fear of punishment therefor.
Whether or not any effects of value were left.
Hour and date and place or port of desertion.

(2) The entries "Not known" and "Unknown" shall not be inserted in the service-record book until every reasonable effort has been made to discover the probable cause of desertion.

707

Report of apprehension, etc.—Whenever a deserter surrenders or is apprehended, careful inquiry shall be made to ascertain the true cause leading up to his desertion and report made to Headquarters, the report to include statements in regard to the following points:

The manner of his return, whether forced or voluntary, and if voluntary, whether through fear of arrest.

The physical condition of the person so returned, as stated by the medical officer.

Statements made to prove identity.

Statements on surrender as to whether he is a deserter or a straggler.

Whether in plain clothes or in uniform upon return.

Condition of clothing, whether adequate or not for the season and part of the country from which returned, and whether ragged or not.

708

(1) *Payment of reward.*—The officer receiving a deserter shall satisfy himself that the man is a deserter and that he is the person he is represented to be, and, if any doubt exists with respect to his identity, shall telegraph his personal description and statement of service claimed to Headquarters, with a request for verification.

(2) When the identity of a deserter is established, the commanding officer of the post shall prepare the necessary vouchers for the payment of the reward and forward them to the nearest disbursing officer of the quartermaster's department for payment. Whenever practicable, a copy of the offer of reward, including the description of the man, will accompany the voucher. The disbursing officer of the quartermaster's department receiving such voucher shall, if the same be in due form, make the necessary payment.

709

Deserters shall be taken up for pay on the rolls of the post designated by the Major General Commandant. Checkages of the amount of the reward offered shall be noted on said pay rolls, a certificate to this effect being made on the face of the vouchers covering the reward before their transmittal to the disbursing officer of the quartermaster's department. Checkages shall also be noted on the pay rolls of all amounts expended by the Government incident to the apprehension and return to the barracks of the deserter, including medical examination, transportation, subsistence, telegrams, etc., and the officer noting these checkages on the pay rolls shall notify the quartermaster, to the end that the proper appropriations may be reimbursed.

710

Physical examination.—The officer receiving a deserter shall also cause him to be examined by a medical officer, and if he be found physically unfit will so telegraph to Headquarters. By physical unfitness of a deserter is meant that degree of unfitness resulting from disease or disability which would render him, in case of his return to the Marine Corps authorities, a menace to the health of those with whom he would come in contact, or which would put the Government to a greater expense in his care and treatment than would be warranted by the benefit accruing to the service by reason of his punishment. The test of unfitness in the case of a deserter is entirely different from the standard set in the case of an applicant for enlistment. While such disabilities as loss of fingers, flat feet, underweight, impaired vision or hearing, loss of teeth, varicose veins, etc., would cause the rejection of an applicant for enlistment, their existence in a deserter will not render him unfit and should not constitute a reason why he should escape punishment. To constitute unfitness in a deserter there must exist disabilities or diseases of a more serious and vital nature, such as insanity, tuberculosis, appendicitis, diseases of a contagious nature, etc. Again, a distinction should be made between deserters for whom a reward has been paid for delivery and those who voluntarily surrender; in the former case, i. e., where a reward has been paid for delivery, the deserter will not be pronounced "unfit" except as above outlined, while in the latter case, where the deserter voluntarily surrenders, he will be pronounced "unfit" if the disability be such as would unquestionably preclude enlistment. In either case, the medical examiner should assure himself that the man is not feigning disability.

711

Report of delivery, etc., made to Headquarters.—When an enlisted man who has been absent without or over leave for any cause for a greater period than 10 days surrenders or is delivered at a post, a full report on the case shall be made to Headquarters, accompanied by a written statement by the man, a list of witnesses, and, if practicable, statements of his accounts from the date of delivery or surrender to the date of report. No disciplinary action in the case shall be taken until instructions from Headquarters are received. If a reward has been paid for delivery, this fact and the amount will be stated in this report and the amount will be deducted in the statement of pay account.

712

Discharge as unfit.—When a deserter or a straggler is discharged as unfit, an itemized statement of all expenses incurred will be reported to the paymaster and the quartermaster furnished with a certificate that the necessary checkage has been made.

713

(1) *Discouragement of straggling.*—Every effort will be made to discourage the practice of marines willfully absenting themselves and reporting in at other stations for duty.

(2) *Stragglers and deserters will be returned to their proper stations or ships,* whenever practicable, without specific directions from Headquarters, except in cases where an inordinate expense will be incurred, and in which cases instructions will be requested. Unless assured that travel will be performed voluntarily, in which cases transportation may be furnished, absentees will be returned to their stations or ships under guard. Government transportation will always be utilized when available; requests for transportation on naval vessels to be made upon the local naval authorities. Statements of all expenses incurred will be forwarded to the proper officers, with request for checkage against the pay accounts of the marines concerned, but commanding officers will in all cases exercise discretion, to the end that no needless or extraordinary expense, with resultant checkage, is incurred.

(3) *Desertion in time of war.*—As marines who deserted in time of war and who have been in desertion more than two years are no longer amenable to naval jurisdiction, care will be exercised to establish the identity of an alleged apprehended deserter or straggler and his present status before taking further action.

714

(1) *Man's commanding officer notified by telegraph.*—When a straggler from another post is apprehended or surrenders within 10 days from the time of his original absence, the commanding officer will immediately telegraph this in-

formation to the commanding officer of the barracks from which said man is absent, giving the date and hour of his original surrender.

(2) Upon the receipt of such a telegram the commanding officer shall retain the staff returns until orders as to their disposition are received from Headquarters.

715

Applicants eloping, etc.—When an applicant for enlistment transferred from a recruiting station to a barracks elopes, refuses to complete enlistment, or is rejected, report will be made to the officer in charge of recruiting direct. A copy of the report will be sent to the recruiting officer who accepted the applicant, and (in case of rejection by the medical officer) a copy to the Bureau of Medicine and Surgery.

Deck and Summary Court-Martial Memoranda.

716

Who makes, where sent.—Commanding officers of marines at shore stations shall prepare the deck and summary court-martial memoranda in triplicate on forms N. M. C. 512, 512a, and 512b. One copy shall be transmitted to the Adjutant and Inspector, Marine Corps, one copy mailed direct to the Auditor for the Navy Department, and the third copy pasted in service-record book.

717

(1) *Absence without leave and over leave.*—Care shall be taken in describing the offenses of absence without leave and absence over leave. The memoranda shall show briefly, in substantial accordance with the following example, the dates and the hours of the beginning and the ending of the unauthorized absence:

"AWOL (or AOL) 7.30 a. m., March 14, 1916, to 7 p. m., March 16, 1916, when reported at ———"

Absence or return beginning or ending at noon or midnight should be stated "12 noon" or "12 midnight."

(2) If the man surrendered at a post or station other than that from which he absented himself, and was returned to the latter, the memorandum shall contain additional data similar to the following:

"Joined this command therefrom, March 18, 1916."

(3) All entries under "Remarks" on pay rolls and muster rolls shall agree with the data contained in the memorandum above mentioned.

(4) *Loss of pay.*—Commanding officers of marines at shore stations, and other officers in command of detachments keeping and rendering pay rolls to the paymaster's department, United States Marine Corps, will note, over their signatures, on the records of summary and deck courts, that the loss of pay in the particular case, if any has been adjudged and approved, has been noted in service-record book.

General Courts-Martial.

718

Recommendations for trial by general courts-martial should be so complete of themselves as to afford all important facts necessary for the preparation of charges and specifications. Service-record books will, whenever available, accompany recommendations for trial for unauthorized absence. In cases of theft, the articles stolen, ownership, place and time of theft, the value of items will be clearly specified. Original checks will, whenever practicable, accompany recommendations for trial for unlawful check transactions. (Photographic copies will be accepted when originals can not be produced.) While the Navy Regulations do not require the furnishing of specimen charges and specifications, the receipt of same often proves of considerable aid and value as indicating precisely the real nature and extent of the offense charged and to the allegations of which the offender would be held to confess should he plead guilty.

719

It is the policy of the Department not to resort to trial by general court-martial except in those cases where it is clearly evident that a deck court or a summary court-martial would be unable to award a suitable and effective punishment.

Arrangement of Men in Conduct Classes.

720

(1) Enlisted men of the Marine Corps, serving on shore, shall be arranged by their commanding officer, in order of good conduct, in four classes, namely, first, second, third, and fourth.

(2) When such classification is first made, preference shall be given to men with good records and of long standing in the service, and such changes in classification shall, from time to time, be made as may be warranted by the conduct of the men.

(3) First-class-conduct men shall be granted every privilege consistent with discipline and the demands of duty. From them shall be formed a special class of men upon whom full reliance may be placed.

(4) For men in lower classes such restrictions shall be established by the commanding officer as he may deem proper.

CHAPTER 8.

TRANSFERS.

General Instructions.

801

(1) *Staff returns.*—When a marine is transferred from one station to another, the officer transferring him shall, at the same time, forward his staff returns to the officer to whom he is transferred, including service-record book signed by himself, which shall, under the proper heading, contain the dates of promotions and reductions with the reasons for the latter; the dates of beginning and ending of any special details as cook, or messman; the date of departure from or arrival within the United States, in going to or returning from foreign shore service; the record of last settlement or payment of the marine, including balances, if any, as shown by the last pay roll as audited and settled by the officer carrying the accounts of a ship or station; the amount of indebtedness to the post exchange; and such other information as may be known to him concerning the man's military history; also a complete list of offenses committed and punishments awarded at the station or on board the ship from which he is transferred, so that a continuous record of the soldier's conduct during current enlistment may at all times be in the possession of his commanding officer.

(2) *The service-record book* shall be kept in accordance with the instructions contained therein, all entries being made as they occur; and upon transfer shall be completed and signed by the officer transferring the man. The marks in conduct on transfer shall correspond with the record of the man as shown by the punishments adjudged.

(3) *The transfer of marines from a ship to a hospital*, and their discharge therefrom shall be governed by the same rules as are provided in article 1203, Navy Regulations, 1920, for enlisted men of the Navy, substituting, where necessary, "Major General Commandant" for "Bureau of Navigation," and "marine barracks" for "receiving ship."

(4) *Discharges, bad conduct and dishonorable, at Mare Island.*—When a marine is transferred from any station outside the continental limits of the United States to the marine barracks, Mare Island, Calif., for bad conduct or dishonorable discharge in accordance with the sentence of either a summary or a general court-martial, all business pertaining to such discharge shall be transacted as expeditiously as possible and in accordance with such procedure as may be prescribed by the Major General Commandant.

(5) *When ordered to detail a detachment for service on board ship*, the commanding officer of marines shall carefully select men of the best character for such duty, and shall make such selection without unnecessary delay, in order that they may have time for preparation. Men having less than two years to serve shall not be detailed for duty on board a vessel destined to a foreign station; nor, except in cases of emergency, shall recruits be detailed for service afloat unless they have been thoroughly instructed in regard to their duties on board ship.

802

Outfit of uniform.—Men shall not be transferred from a shore station to a seagoing ship for duty without a complete outfit of uniform.

803

Deficiencies in the complement of marines on board ships on the eve of sailing may, by the order of the commandant of the station, be supplied by the commanding officer of marines, who shall report the circumstances to the Major General Commandant without delay.

804

(1) *Telegraphic notification of time of arrival.*—Upon transfer of detachments of more than five men the post to which transferred will be notified, by letter or telegram, a reasonable time in advance of arrival, giving the number of men in the detachment, the date, hour, and place of probable arrival.

(2) *Transfer by staff returns.*—When enlisted men are transferred by staff returns the post or vessel to which the transfers are made will be notified as to the time at which the men may be expected to arrive, and also as to the cause of any delay that may be anticipated.

(3) When men are transferred, singly or in groups, and their staff returns (service-record books) are transmitted to the new station by mail, a carbon copy of the orders (or a memorandum) covering such cases, showing date and place from and to which transferred, with notation of any furlough or delay granted en route, shall be attached to the outside of each service-record book.

805

Staff returns; to whom sent.—Upon transfer of marines to a post for duty with detachments or companies stationed or being organized thereat, the outside envelope or wrapper containing their staff returns will be in every case addressed to the commanding officer of the post to which transferred. When the staff returns of men are intended for detachment or company commanders this may be indicated by placing them in inner envelopes or wrappers marked with designation of the detachment or company.

806

Restored probationers or former general court-martial prisoners shall not be transferred for duty as prison guards except in case of emergency and upon specific orders from Headquarters. They shall not be transferred from one station to another to the exclusion of other enlisted men who are eligible for transfer but whose retention is desired in preference.

Army Transports.

807

(1) *Smallpox.*—Every detachment transferred from a post in the United States for embarkation in a United States Army transport will be provided with the certificate of a naval surgeon to the effect that its members have been duly inspected and are protected against smallpox.

(2) *Epidemic disease.*—In case an epidemic disease appears in any detachment of marines en route for embarkation on an Army transport, the commanding officer of the detachment will at once notify the senior Army surgeon at the

place of sailing by telegraph. The commanding officer of the post from which the detachment leaves will direct the commanding officer of the detachment to carry this provision into effect should occasion arise.

808

(1) *Orders for transfer by transport.*—The officer transferring a draft or detachment of enlisted men for embarkation on an Army transport will furnish the officer or enlisted man in charge with an order directing him to report to the transport quartermaster, and in all cases there will be included in the order a designation of the messes in which the men comprising the draft or detachment are entitled to be subsisted.

(2) *Messes.*—The following provisions are in accordance with the Transport Regulations, United States Army, and will govern so far as they pertain to the United States Marine Corps:

(*a*) Ship's officers' mess: Sergeants major, quartermaster sergeants, first sergeants, and gunnery sergeants.

(*b*) Troop mess: All other enlisted men of the Marine Corps, including prisoners, of whatever rank.

(*c*) Hospital mess: All sick men of the Marine Corps, irrespective of rank, requiring special diet.

Sea and Foreign Service.

809

It is imperative that, as far as practicable, the sea and foreign service of enlisted men of the Marine Corps be equalized. Commanding marine officers when detailing men for foreign service will not, without specific authority from the Major General Commandant, detail any man who has not at least 15 months to serve on his current enlistment, nor, in the case of reenlisted men, any man who within the preceding 12 months completed 15 months' sea or foreign service.

Officers' Choice of Station.

810

Officers are afforded the opportunity, through the medium of official letters to the Major General Commandant, to indicate preference for duty and station prior to probable date of change of station.

CHAPTER 9.

RECORDS, REPORTS, ETC.

Muster Rolls.

901

(1) *When required.*—A muster roll is required from each staff office, depot of supplies, barracks detachment, marine detachment of a ship, hospital, or legation, field and staff and headquarters detachment of a brigade, regiment, or other organization, company, and recruiting division for every month or fractional part thereof, if only one day, during which marines have been attached thereto; also from every officer outside of Washington who is not attached to a station, a recruiting district, or carried in the Headquarters muster roll as in hospital or on sick leave, but who is, for example, on duty at the Naval War College, Newport, R. I., the Army Service Schools, Fort Leavenworth, Kans., as attaché at Peking, as fleet marine officer not in command of marine detachment, etc., the roll to be rendered by the senior where two or more officers are serving together under these conditions; also from enlisted men when, under exceptional circumstances they are separated from their organization for the performance of special duties. The roll will be prepared to include the last day of the calendar month and forwarded as soon thereafter as practicable, and must not cover a period involving parts of two months (see paragraph 903). When a new organization is formed, a roster of same will be immediately forwarded to the Adjutant and Inspector. The authority for the organization of a company, detachment, etc., shall be entered on sheet 1, below the recapitulation.

(2) *Complete rolls and supplemental rolls.*—The rolls for January, April, July, and October will be complete (see par. 902). Rolls for the other months of the year will be supplemental and will contain only such facts as may be necessary to complete the military history of the marine. Names of officers and men will not be repeated on the supplemental rolls if they appeared on the previous roll of the quarter unless additional remarks are necessary. The entry of the date of enlistment, except in cases of men joining, transferred, discharged, died, or deserted, should not appear on any but the first roll of the quarter. Remarks such as "SD clerk," "Det. d. Radio Station," etc., need not be repeated on supplemental rolls for the quarter.

902

When a station is abandoned, a ship placed out of commission, or any organization disbanded, a complete roll will be prepared and forwarded immediately, the final disposition of the personnel being shown thereon in addition to the usual remarks.

903

Detachments en route from the United States to foreign stations, or returning therefrom, will render a muster roll on commencement and completion of the voyage, giving the name of the vessel on which passage was taken, the date of joining, the organization from which joined, and the date and port from which sailed.

904

(1) *Officers ordered to naval hospitals.*—When a marine officer is detached from duty in Guam, the Philippines, China, or Japan, or from a ship of the Pacific or Asiatic Fleet, and ordered to the naval hospital, Mare Island, Calif., he will immediately after admission thereto report by letter, through official channels, to the commanding officer, marine barracks, Mare Island, Calif., stating the station from which detached and date of detachment, the authority, date of arrival in San Francisco, conveyance, and date of admission to hospital. Upon receipt of such a letter, the commanding officer, marine barracks, Mare Island, Calif., will take up on his muster roll said officer, with remarks that he reported by letter ———, 19——, from (name of post or ship), having arrived at San Francisco, Calif., ———, 19——, via (name of vessel), and having been admitted to naval hospital, Mare Island, Calif., ———, 19——. He will continue to carry said officer on the roll, whether in hospital or on sick leave, until detached by the Major General Commandant.

(2) The same procedure will be followed when an officer is detached from duty in the Dominican Republic or the Republic of Haiti or other foreign station, and ordered to a naval hospital in the United States other than the naval hospital, Washington, D. C., substituting in place of the Mare Island hospital and barracks the proper hospital and barracks.

(3) Officers who are patients at any United States Veterans' hospital will be carried on the rolls of the organizations to which they were attached when admitted to such hospital, unless orders for their detachment are received.

905

Naval Hospital, Washington, D. C.—When an officer is detached from a post or ship, and ordered to Washington, D. C., for admission to the United States naval hospital there, he will report, as above directed, to the Major General Commandant and will be taken up on the Headquarters Marine Corps muster roll.

906

Leave or delay en route.—When an officer who has been detached from one station and ordered to another is granted leave of absence or authorized to delay reporting at his new station, he will, on reporting for duty, inform his commanding officer of the inclusive dates of leave or of delay of which he may have availed himself. The delay, which is to be reported on the muster roll, shall exclude the actual travel time, the four days allowed by orders which do not express haste, and the date of reporting for duty (see arts. 132 and 1727, N. R.). This information will be embodied in the remarks immediately after the entry of the fact of the officer's joining on the first muster roll on which his name appears at his new station.

907

(1) *Absence—Own misconduct.*—Article 554, Navy Regulations, regarding forfeiture of pay when an officer or enlisted man of the Marine Corps is absent from duty on account of injury, sickness, or disease resulting from his own intemperate use of drugs or alcoholic liquors or other misconduct requires an appropriate entry under "Remarks" on muster rolls. An entry will be made

on supplementary as well as on complete muster rolls in every case of absence of an officer or enlisted man from duty on account of injury, sickness, or disease, whether due to his own misconduct or not, and the entry will be followed by the words "Art. 554, N. R., applies," or "Art. 554 does not apply," as may be appropriate. Example: 1 to 15, sick, present, Art. 554, N. R., does not apply. The wording "applies" or "does not apply" should invariably be used, owing to frequent omissions through inadvertence of the word "not" from the unauthorized entry "does apply."

(2) When the injury, sickness, or disease begins in one month and terminates in another and is continuous, the entry on the first roll as to whether or not article 554, N. R., applies will not be changed on subsequent rolls unless the reason for such change is shown; for example, "Change of diagnosis."

908

General court-martial prisoners.—An entry will be made opposite the name of each general court-martial prisoner on the muster rolls to show whether his sentence included dishonorable discharge or not, as follows: "DD" or "not DD," as the case may be, and the number of general court-martial prisoners, DD, will be shown on the recapitulation in the proper place. No one will be shown as a general court-martial prisoner until his sentence has been approved.

909

How written.—When practicable, muster rolls will be typewritten, black record ribbon being used, and all entries on a roll made with the same machine. When hand written, black ink will be used and all entries made by the same person in a clear and legible hand. The roll must be free from erasures and interlineations. No blank lines should be left between entries nor between the last entry in the column of remarks and the signature to the roll.

910

Sheets, how headed.—Each sheet will be headed by the full name of the organization (company, battalion, regiment, and brigade), location, and dates, and will be consecutively numbered.

911

Entries will be arranged in the following order:

(1) In the body of the January, April, July, and October rolls the names of all officers and enlisted men attached to the command on the last day of the period for which the roll is rendered, including stragglers and deserters who have surrendered or been delivered and taken up on the strength of the command; in supplemental rolls, as necessary.

(2) Under separate headings, in the following order: The names of all officers and enlisted men who were detached, transferred, retired, resigned, dismissed, accounts closed to accept warrant or commission, discharged, died, deserted, enlistment canceled, during the period for which the roll is rendered. The regular Marine Corps personnel present, transferred, etc., shall appear first on the roll, followed in similar order of headings by the personnel of the Fleet Marine Corps Reserve, the Marine Corps Reserve (by classes, lowest numbered class first, omitting subclasses), and other arms of the United States services temporarily serving with the United States Marine Corps. In other

words, each branch of the Marine Corps and each other arm of the service shall be shown entire in a separate and distinct portion of the roll.

(3) Officers and enlisted men on temporary detached duty with an organization shall be carried on the roll of the organization with which temporarily serving under the heading "Temporarily attached" immediately following the last entry in the body of the roll, and shall be shown in the recapitulation as "Temporarily attached" on one of the blank lines provided. They shall also be shown on the roll of the organization to which regularly attached as on temporary detached duty with the temporary organization, and shall be accounted for in the regular manner in the recapitulation.

912

Headings.—The headings "Detached," "Transferred," "Retired," etc., will be entered in the center of the line following the last entry under the preceding heading, and all headings will be underscored in red ink.

913

"No." column.—In the "No." column the names in each grade in the body and under each heading of the roll will be separately numbered consecutively.

914

(1) *"Name" column.*—In the "Name" column the surname will appear first, followed by Christian name in full and initials of middle names, if any. In this column the rank of the officers and men will also be entered, the entry being made on the line preceding the first name in each separate grade. The same person's name shall not appear twice on a roll except in the cases of men whose accounts are closed to accept warrant or commissioned rank and who accept such rank and are shown thereunder on the same roll, and men who are both discharged from and rejoin the organization by reenlistment or transfer during the period covered by the roll. Men who are both transferred and who have rejoined the organization during the period covered by the roll should be shown in the body of the roll only.

(2) *Officers' names.*—In the body of the roll and under separate heading the names of officers will be arranged under each grade, not in the alphabetical sequence of surnames, but according to seniority. The department to which a staff officer belongs will be indicated, not above his name, but opposite his name in the column of remarks.

(3) *Enlisted men's names.*—The names of enlisted men in each grade will be arranged alphabetically, according to the sequence of the letters of the surname, e. g., "Bennet" before "Brown," "Delmont" before "Delmore," etc.

915

Date of enlistment.—In the "Enlisted" column will be entered the date of last enlistment or reenlistment, except when omitted as provided in par. 901 (2). The figures of the day of the month will be entered first, followed by the first three letters of the name of the month and the last two figures of the year, as follows, punctuation marks being omitted:

30 Sep 17
7 Jul 16
16 Mar 18

916

In the "Remarks" column the following data will be entered, in chronological order, opposite the names of the persons concerned (see par. 918):

(1) *Joinings.*—Date joined, from what organization, and, if sea travel is involved, via what vessel, with date and port of departure and date and port of arrival. In cases of men first joining by staff returns, date of receipt of staff returns and subsequent date of joining in person.

(2) *Promotions and reductions.*—Promotions and reductions, with date thereof (not date of receipt). If by other authority than the Major General Commandant, the authority should be stated. In the case of appointment or promotion of officers, the remark should show date of acceptance of appointment, date of execution of oath of office, and the date from which the officer takes rank.

(3) *Duties performed by officers.*

(4) *Special or detached duty details,* specifying the particular kind of duty performed.

(5) *Mail clerks, specialists and special duty.*—When an enlisted man is appointed or relieved as Navy or assistant Navy mail clerk, the fact and date of such appointment or relief. Enlisted men detailed to perform specific services which remove them temporarily, from the ordinary duty roster of the post, detachment, or organization to which they belong, shall be shown on the muster rolls as on duty as specialists, if so rated; otherwise, as on special duty (see par. (7)).

(*a*) The first muster roll rendered after a specialist is rated will show the class, date of rating, and authority for rating; subsequent rolls need show only the class in which rated, as, for example,

"Spl. 3 Cl."

when disrated, the muster roll will show the date of disrating and the authority therefor. Men detailed on special duty will be shown as in the following examples:

"SD., post exchange steward,"
"SD., mail orderly," etc.

(6) Sergeants major, quartermaster sergeants, first sergeants, and gunnery sergeants, when performing duties pertaining to their respective ranks should not be shown in the "Remarks" column as on special duty, but the particular kind of duty should be stated; as for example, "Post," "Brigade," "Regimental" or "Battalion Sergeant Major," "in charge of clothing room," "in charge of arms and accoutrements," etc.

(7) *Gun captains, signalmen, cooks, etc.*—Details as gun captains or gun pointers, with class and kind of gun to which detailed; signalmen, with class; cooks, with class; or messmen. On the roll in which the man's detail first appears, the date of detail, and when revoked the date of revocation, with reason therefor.

(8) Inclusive dates of officers' absence on leave or delay, with authority (see Art. 1727, N. R.).

(9) *Inclusive dates of absence of men on furlough.*—When an officer or enlisted man is granted leave or furlough on foreign station to visit the United States, the name of the conveyance, with the dates of departure and arrival at ports in the United States.

(*a*) NOTE.—If these details, etc., (3 to 9) have been continuous during the whole period covered by the roll, only the nature of the duty, etc., need be shown. If, however, they have not been continuous during the period covered by the roll, the dates inclusive of detail, etc., shall be given. When an officer or man is shown as in a particular duty status for a single day, the word "only" shall follow the date shown.

(10) *Dates covering sick,* present, or in hospital, naming hospital. The date of admission will be shown as a day of sickness, while the date of discharge to duty will be shown as a day of duty. (See Art. 907.)

(11) *Absences without leave or over leave,* giving the hours and dates, the day of the month to follow the hour in every instance, e. g., "AWOL from 8 am 3 to 5 pm 3"; or "AOL from 12 midnight 5 to 3.30 pm 16." The number of days or hours comprising such absence should not be stated. When the absence begins or ends at noon or midnight the terms "12 noon" and "12 midnight" shall be used, not 12 n, 12 m, 12 am nor 12 pm. When a man has been absent over or without leave, but upon return has presented an excuse which was accepted and the absence thereby excused. "Excuse accepted" shall be entered after the entry of the absence. A man's unauthorized absence is terminated upon his surrender or delivery and acceptance by Marine Corps or naval authorities at any station or on board any ship. If he has been in the hands of civil authorities, the entry on the roll shall show whether he was convicted, acquitted, or released without trial, with the nature of the effense, and the sentence imposed, if any; also whether on liberty or furlough at the time of arrest by civil authorities, and when such liberty or furlough expired, since a man arrested and not convicted by the civil authorities is not considered as absent without leave if he was on authorized absence when taken into custody. When not definitely determinable, the hours and dates of the beginning and termination of a man's unauthorized absence shall be recorded as the hours and dates his absence and return became officially known. When a returned absentee claims to have been sick, absent, in addition to the entry of disciplinary action or acceptance of excuse, the muster roll should state whether or not article 554, Navy Regulations, applies.

(12) *Date and nature of any offense,* arrest or confinement and release therefrom, stating whether confinement was "awaiting action," "awaiting trial," "awaiting sentence" or "serving sentence," and dates and nature of all punishments awarded, and for what offenses. In stating the nature of an offense, the use of general terms such as "neglect of duty," "unmilitary conduct," "insubordination" and "violation of post order" should be avoided, care being taken to so describe the acts constituting the offense as to show its seriousness or triviality and to afford the information essential to a proper estimate of the quality of a man's service, of the conduct markings he should receive, and of the character which should be awarded him on discharge. When a man is sentenced to confinement and is at large awaiting an empty cell, or prisoner-at-large for any reason, such fact shall be shown; the roll should show whether or not he was in a full-duty status. In cases of restoration to duty after being confined on suspicion, or pending investigation, or action of higher authority, or after a term of confinement or period of probation imposed by court-martial, the date of release and restoration to duty shall be known. Like entry shall be made in cases of restoration to duty before the expiration of term of confinement imposed by deck court, summary court-martial or otherwise, showing the authority for the remission of the unserved portion of such confinement. When men, not serving sentence of a court, become sick while confined, the roll should show whether or not the confinement continues during sickness. If it does so continue, the period is accounted "time lost" and is not credited to the man as service. If it does not so continue, the period of sickness is not accounted "time lost," unless article 554, Navy Regulations, applies. When a man has been restored to duty on probation, the fact shall be entered on the roll for the month during which he was so restored, and the roll for the month during which the probationary period ends shall show the termination and the date thereof.

(13) *Dates of trial* by deck courts or courts-martial, sentences, and the action of the convening authority on deck courts and general courts-martial, and of the immediate superior in command on summary courts martial, with date of such action. Where loss of pay adjudged is remitted subject to conditions specified

in article 1877, Navy Regulations, the roll should state "LP remitted, Art. 1877." Particular care should be exercised in the entry of court-martial data, as a variance between the roll and the court-martial memorandum necessitates correspondence and correction of the record in error. Only pertinent facts should be shown regarding courts. Charges and specifications need not appear.

Example 1: 1 to 5, on furlough; AOL from 12 midnight 6 to 5.30 am 15; for which tried by SCM 16, and sentenced as mitigated to lose $10, LP remitted, art. 1877; 18 app by ISIC; confined 15 awtg trial and 16 to 18 awtg result of SCM; 19 restored to duty.

Example 2: 1 to 15, conf awtg trial by GCM; 16, tried by GCM found guilty of AWOL (see Feb. roll for absence) and sentenced as mitigated to 1 year conf at hard labor and to lose all pay except $3 per mo; 16 to 28 conf awtg result of trial; 29, sentence app by CA; 29 to 31, conf awtg transfer (or, 30 to N.P., Portsmouth, N. H., a GCMP)

(14) ***Bestowal of medals,*** bars, pins, badges, certificates, with serial number of each; commemorative ribbons; and letters of commendation, the service for which awarded and the date of delivery. (See art. 523.)

(15) ***Date and nature of any battle,*** expedition, affair, or skirmish participated in by any officer or enlisted man, also date and nature of any injury received, during the period for which the roll is rendered. (See art. 918.)

(16) ***All details for temporary duty on shore*** by marines attached to vessels, with nature and period thereof, arranged in sequence as to date. (See art. 918.)

(17) ***The date and station or ship to which detached or transferred;*** and when detached or transferred to or from a point outside the limits of the United States, the name of the conveyance, port, and date of departure. (See 916 (1).) Those transferred as general court-martial prisoners will be so shown. When men are actually transferred from ships' detachments or other organizations to hospitals for treatment, the rolls will show the place to which their staff returns are forwarded. Remarks showing transfer to inactive status shall specify the reserve division to which staff-returns are transferred.

(18) ***Retirement.***—Date, cause of retirement, and, in the case of an enlisted man, the character that would have been awarded had he been discharged.

(19) ***Resignation.***—Date acceptance of resignation takes effect, also date received by officer, if subsequent.

(20) ***Dismissal.***—Date and authority, also date notice of dismissal received by officer.

(21) ***Discharge.***—The date and cause of discharge as given by discharge certificate, character given on discharge, and cause of retention if retained beyond expiration of enlistment. Where there is but one cause for retention, the period of retention should not be shown, as it can readily be ascertained from the dates of enlistment and discharge; where there is more than one cause of retention, the number of days retained for each cause should be stated. Men accepting appointment to warrant or commissioned rank will be shown under the heading "ACCOUNTS CLOSED TO ACCEPT WARRANT (or COMMISSION)", as may be appropriate.

(22) ***Death.***—Date, place, and cause of death, whether or not in line of duty, whether article 554, Navy Regulations, applies, and character that would have been awarded had the man been discharged. The date and place of burial (giving number of grave, section, etc.) or disposition of body.

(23) ***Desertion.***—Date and hour of desertion, date of desertion being the first day of absence over or without leave.

(24) ***Enlistment canceled.***—The cancellation of an enlistment dates from date of receipt of the order of cancellation. The date of the authority or order therefor shall be entered, with other appropriate remarks.

(25) ***Ditto marks,*** symbols, the remarks "Same as above," and other remarks signifying repetition of remarks in another entry will not be used except when special authority for their use on a particular roll is conferred by letter from Headquarters, Marine Corps.

917

Abbreviations.—The use of the following abbreviations is authorized in the "Remarks" column:

AOL	Absent over leave.
AWOL	Absent without leave.
Adj	Adjutant.
A&I	Adjutant and Inspector.
A&ID	Adjutant and Inspector's department.
App	Approval, approved, appointed.
Arty	Artillery.
Auth	Authority.
Awtg	Awaiting.
BCD	Bad-conduct discharge.
Btry	Battery.
Bn	Battalion.
B&W	Bread and water (solitary confinement).
Brig	Brigade.
Brig. Gen	Brigadier General.
Capt	Captain.
Col	Colonel.
C. in C	Commander in chief.
CO	Commanding officer.
Co	Company.
Con. serv. sen	Confined serving sentence.
Con. tem. reg	Confirmed temporary regular.
Con. tem. tech	Confirmed temporary technical.
CA	Convening authority, civil authorities.
Cpl	Corporal.
DC	Deck court.
Det. d	Detached duty.
DD	Dishonorably discharged; dishonorable discharge.
Div	Division.
Dmr	Drummer.
ER	Expert rifleman.
EPD	Extra police duty.
1st Lt	First Lieutenant.
1st Sgt	First sergeant.
GCM	General court-martial.
GCMP	General court-martial prisoner.
Gy. Sgt	Gunnery sergeant.
HQMC	Headquarters, Marine Corps.
IHCA	In hands of civil authorities.
ISIC	Immediate superior in command.
Jd. fr	Joined from.
JAG	Judge Advocate General.
Lt. Col	Lieutenant Colonel.
Maj	Major.
Maj. Gen	Major General.
MGC	Major General Commandant.
MB, NS	Marine Barracks, Naval Station.
MB, NYD	Marine Barracks, navy yard.
M. Gun	Marine gunner.
Mm	Marksman.
MS	Medical survey.
Mus. 1cl	Musician, first class.
Mus. 2cl	Musician, second class.
Mus. 3cl	Musician, third class.

ND	Navy Department.
OIC	Officer in charge.
P. wrnt	Permanent warrant.
PE	Post exchange.
PClk	Pay clerk.
Plat	Platoon.
PQM	Post quartermaster.
Prin. mus	Principal musician.
PAL	Prisoner-at-large.
Pvt	Private.
Pvt. 1cl	Private, first class.
Prob. reg	Probationary regular.
Prob. tech	Probationary technical.
QMClk	Quartermaster clerk.
QMSgt	Quartermaster sergeant.
R. dep	Recruit depot.
R. dis	Recruit district.
RO	Recruiting officer.
RS	Recruiting station.
R. Wrnt	Recruiting warrant.
Red	Reduced.
2d Lt	Second Lieutenant.
SGP, 1cl	Secondary gun pointer, first class.
SGP, 2cl	Secondary gun pointer, second class.
Sgt	Sergeant.
Sgt. Maj	Sergeant major.
SOP	Senior officer present.
Spl. 1cl	Specialist, first class.
Spl. 2cl	Specialist, second class.
Spl. 3cl	Specialist, third class.
Spl. 4cl	Specialist, fourth class.
Spl. 5cl	Specialist, fifth class.
Ss	Sharpshooter.
S. Wrnt	Ship's warrant.
Sig. 1cl	Signalman, first class.
Sig. 2cl	Signalman, second class.
SD	Special duty.
Sq	Squadron.
Sqd	Squad.
S/R	Staff returns.
SCM	Summary court-martial.
Tech. Wrnt	Technical warrant.
Tpr	Trumpeter.
T. Wrnt	Temporary warrant.
USMC	United States Marine Corps.

918

(1) *Change of station.*—The fact that the command or company has changed station or designation, with dates, will be shown on sheet 1; for example, when a company leaves a post for expeditionary service, separate rolls for parts of the month will not be rendered, but a roll covering the full month prepared; the note on sheet 1 will show the dates at the station from which transferred, dates of embarkation and debarkation, dates en route, and dates at new station during the month. Whenever any movement of the command, company, etc., occurs in a month other than January, April, July, or October, and a remark in blanket form is attached to the roll covering the same, *the name of every member of the command, company, etc., must appear on such roll;* and whenever remarks in blanket form cover the entire personnel, *the names of such personnel must appear on the roll.*

(2) Under these circumstances the company officers, if not detached from the post, will also be carried on the roll of the barracks detachment of the post from which temporarily absent. In the roll of a detachment on board a cruising vessel the location of the vessel during the period covered by the roll will be shown on sheet 1. When all members of an organization are transferred from the same post or ship to another post or ship, the muster roll of the post or ship from which transferred will show the date of transfer and the post or ship to which transferred; and the muster roll of the post or ship they join will show the date of joining and the post or ship from which joined, in each case by a note immediately after the last entry in the roll and above the signature of the officer or noncommissioned officer rendering the roll. Similarly, if all the members of the command have been on detached service during the whole or any part of the period for which the roll is rendered, like notation will be made at the foot of the roll. In cases where the same remark applies to all but a few of the command, like notation may be made, the note reciting the fact that the remark applies to all members of the detachment except (naming persons excepted).

919

Recapitulation.—In the "Recapitulation", printed on sheet 1, will be shown the distribution of the command on the last day of the period for which the roll is rendered, and the losses during that period. Men of the Fleet Marine Corps Reserve, Marine Corps Reserve, Navy personnel and other arms of the service will be shown under proper headings and on the provided or blank lines of the recapitulation. At posts where there are brigade, regimental, company, etc., organizations, in addition to the recapitulation for each separate command, the commanding officer will render a general recapitulation comprising the entire command.

920

Signature.—The roll shall be signed, immediately after the last entry therein, by the commanding officer as "Commanding"; or, in case there is no officer present by the senior noncommissioned officer as "In charge," adding "In the absence of the commanding officer" when appropriate; and by the officer in charge of a recruiting division as "In charge"; the rank of the person signing being given in each case immediately above the designation here indicated.

921

Forwarding.—Muster rolls of stations or detachments on the Pacific coast, the Pacific and Asiatic Fleets, Pearl Harbor, Hawaii, Guam, the Philippines, and Peking, China, shall be transmitted through the assistant adjutant and inspector, San Francisco, Calif. Rolls of all other stations or detachments shall be forwarded direct to the Adjutant and Inspector. In the lower left-hand corner of the signature sheet shall be noted the date on which forwarded, and in case of detachments on board ship, the place from which forwarded. The roll must not be cut or mutilated, or fastened together by pasting or any kind of adhesive material, or rolled, but must be forwarded flat.

922

Roll should be complete.—Since the muster roll is the record to which reference is made in the adjustment of claims and other questions affecting those whose names are borne thereon, in its preparation nothing should be omitted which properly pertains to the complete military history of an officer or enlisted man, and in case of doubt an entry will always be preferable to an omission, these instructions covering only the more usual entries required and not precluding proper entries of other matters not specified herein.

923

REPORT OF TRANSFERS AND DISCHARGES.

(1) *When made.*—Report of transfers and discharges of enlisted men will be made to the Adjutant and Inspector as they occur, by the commanding officer of marines.

(2) *Additional entries.*—The following entries will be made on the report of transfers and discharges opposite the name of each man discharged:

(*a*) Character awarded on discharge.

(*b*) A notation of award (if any) of either a good-conduct medal or a good-conduct-medal bar.

(*c*) The address after discharge.

(3) The officer or noncommissioned officer making these reports will forward copies direct to the Quartermaster, the Paymaster, and the pay officer of the ship or station from which transfers are made.

924

(1) *Pay rolls.*—Commanding officers of marines, or other officers charged with the rendition of pay rolls at shore stations, will retain the third copy of the pay roll until the preparation of the pay roll for the next succeeding month shall have been completed. The third copy of the pay roll for the preceding month shall then be forwarded immediately, on the first of the month, when the current roll is sent in for audit and settlement, to the pay office of the ship or station where the original roll was paid.

(2) Commanding officers of marines afloat will forward promptly the third copy of their pay rolls in the manner provided in article 617, Navy Regulations, 1920.

REPORTS ON FITNESS OF OFFICERS.

925

(1) *Preparation.*—In order that reports on fitness of officers may reach headquarters promptly and cover all periods of service, the following instructions regarding the preparation of reports required to be rendered at times other than March 31 and September 30 of each year are issued:

(*a*) Every officer, upon receipt of orders detaching him permanently, or ordering him upon expeditionary service, from the command of a senior, shall submit to such commanding officer, either before leaving or as soon as practicable after leaving his station, a report on fitness with such blanks filled by himself as are designed for that purpose.

(*b*) It shall be the duty of every commanding marine officer, upon receipt by him of orders detaching him permanently, or ordering him upon expeditionary duty, or upon being superseded by a senior, to require all officers under his command to submit to him reports on fitness of themselves filled out as prescribed in the preceding paragraph.

(*c*) In the case of an officer absent from his post upon the detachment or relief of the commanding officer, it shall be his duty, immediately upon his return, to forward to the outgoing commanding officer a report on fitness filled out as above prescribed.

(2) Every officer assigned to duty upon expeditionary service under command of a senior whose duty it is to submit reports on fitness upon him shall, upon being relieved from duty under that senior, submit to said senior, before leaving his command, a report filled out as above prescribed.

(3) When a commanding marine officer is absent from his post on temporary duty other than expeditionary, on leave, or on account of sickness for a longer period than six weeks, he shall exclude from his reports on officers attached to his post the period of such absence; and it shall be the duty of the officer in

temporary command to render reports covering said period upon the resumption of command by the regular commanding officer.

(4) When an officer is absent from his command on temporary duty other than expeditionary, he will, if such duty is continuous for a period of more than six weeks, be reported on during said period by the immediate temporary commanding officer authorized to report upon the fitness of officers.

(5) Upon receipt by an officer of orders to appear for examination preliminary to promotion, he will immediately submit to his commanding officer a report properly prepared and covering the period since last reported on to date of leaving in obedience to said order.

926

(1) *Probationary officers.*—All reports on fitness of officers of the Marine Corps holding probationary appointments for a period of two years will contain a statement under the heading "Remarks," whether or not the officer reported upon is recommended for retention in the service.

(2) In the event that the reporting officer considers that the probationary appointment should be terminated, he will state his reasons therefor in full and will refer the entire report to the officer concerned for statement, which statement will be forwarded to Headquarters with the report.

(3) The above does not preclude commanding officers from recommending at any time the revocation of a probationary appointment, but before forwarding such a recommendation a copy will be furnished the officer concerned for statement, and such statement shall in every case accompany the recommendation for revocation of appointment.

Assistant Adjutant and Inspector, San Francisco.

927

(1) *Jurisdiction.*—The assistant adjutant and inspector, San Francisco, Calif., will have supervision of all reports, returns, muster rolls, staff returns, court-martial memoranda, etc., which are required to be submitted to the Major General Commandant, or to the Adjutant and Inspector, by the following posts and detachments.

Puget Sound, Wash.: Marine barracks.
Mare Island, Calif.: Marine barracks; naval prison.
San Francisco, Calif.: Depot of supplies; assistant paymaster's office; Western recruiting division.
San Diego, Calif.: Marine barracks.
Pearl Harbor, Hawaii: Marine barracks.
Guam: Marine barracks.
Philippines: Marine barracks, Cavite and Olongapo.
Peking, China: Marine detachment.
Detachments serving on board vessels of the Pacific and Asiatic Fleets.

Commanding officers of ships are requested and commanding officers of posts and detachments are directed to forward such papers via the assistant adjutant and inspector, United States Marine Corps,, 36 Annie Street, San Francisco. Such papers will be carefully examined by that officer, and should any errors not of a trivial nature be discovered therein they will be returned for correction. When they have been properly prepared and the necessary information has been compiled therefrom for the use of the assistant adjutant and inspector, San Francisco, they will be forwarded without delay to the proper office.

(2) Requisitions for blank forms, books, and other articles supplied by the adjutant and inspector's department will be made by the above-mentioned posts and detachments on the assistant adjutant and inspector, San Francisco.

929

Register of punishments.—The commanding officer of marines shall cause a register of all punishments inflicted by him to be kept.

930

Punishments inflicted upon commissioned officers shall be reported without delay to the commandant of the station and to the Major General Commandant of the Corps.

VERIFYING AND CERTIFYING CASH ON HAND.

931

(1) The attention of all commanding officers of the Marine Corps under whom disbursing officers of the paymaster's department or their deputies are serving is called to the requirements of paragraphs 36 and 37 of Treasury Department Circular No. 52 of 1907, reading as follows:

"36. Whenever feasible, administrative officers should require disbursing officers under them at the close of business on the last day of periods for which they are required to render accounts to count and schedule, in the presence of a duly authorized and disinterested witness or witnesses, all items of cash, i. e., currency, memorandum payments, and other items to appear in their analyses of balances for which vouchers are not to be submitted to the auditors of the Treasury Department with the current account.

"37. Such duly authorized witnesses should verify the counts and schedules provided for by the preceding paragraph and certify to such fact on the account current."

(2) The foregoing regulations of the Treasury Department must be strictly adhered to, and it shall be the duty of the commanding officer either to detail a commissioned officer to perform this duty, or, if none be available, to perform it himself, in the cases of all disbursing officers of the paymaster's department or their deputies at the close of business on the last day of each month.

(3) In addition to certifying the fact of verification on the account current, the verifying officer will also fill out in his own handwriting and personally forward to the Paymaster, United States Marine Corps, at Headquarters, the certificate contained on N. M. C. 745. In verifying the cash in hands of deputies, two original signed copies of N. M. C. 745 will be made out and certified, one copy being mailed direct to the Paymaster, United States Marine Corps, at Headquarters, and the other mailed direct to the disbursing officer whom the deputy represents.

(4) Disbursing officers and deputies acting independently shall themselves take the necessary steps to have their balances verified and certified by a disinterested officer as above indicated. Form N. M. C. 745 should be made out and forwarded regularly each month in all cases, whether there be any cash on hand or not.

REPORTS OF ARRIVAL AND DEPARTURE OF OFFICERS.

932

(1) Officers will not telegraph to the Major General Commandant or the commanding general Department of the Pacific the information that they have arrived in or departed from the United States, except when the arrival is made on a commercial vessel.

(2) Arrivals and departures of officers will be reported as set forth below:

(*a*) Officers arriving by Government vessel on the east coast will make no report to the Major General Commandant. Such officers will receive their orders for change of station from the commanding officer of the nearest marine barracks.

(*b*) Officers arriving at San Francisco will report in person to the commanding general Department of the Pacific, at 36 Annie Street, San Francisco.

(*c*) Officers traveling on commercial vessels at their own expense by choice, whether on leave or under orders, will upon arrival in the United States at other places than San Francisco report their arrival by prepaid telegram to the Major General Commandant or the commanding general Department of the Pacific, as may be appropriate.

(*d*) Officers ordered to travel by first available transportation will upon arrival in the United States at other places than San Francisco aboard a commercial vessel report by telegram at Government expense to the Major General Commndant or the commanding general Department of the Pacific, as may be appropriate.

(*e*) Officers departing from the United States under orders will, just prior to embarkation, mail Form NMC 332d A&I to the Major General Commandant or the commanding general Department of the Pacific, as may be appropriate.

(3) The commanding officer of troops aboard each transport operating between the West Indies and the United States will immediately upon departure from the last port of call in the West Indies radio the Major General Commandant the names, rank, and status of every Marine officer aboard.

(4) The Commanding General Department of the Pacific will report by telegram to the Major General Commandant the names, rank, and status of all Marine officers reporting their arrival to his department in compliance with this order.

Disposition of Records When Organization Discontinued.

933

(1) When a post or station is abandoned, or an organization, detachment, or guard is disbanded, all records, reports, etc., not otherwise provided for, will be disposed of in the following manner:

(*a*) The files and records of the paymaster's department will be forwarded to the Paymaster, United States Marine Corps, Headquarters, Washington, D. C.; except that those o the west coast will be forwarded to the Assistant Paymaster, United States Marine Corps, San Francisco, Calif.

(*b*) The files and records of the quartermaster's department will be forwarded to the Quartermaster, United States Marine Corps, Headquarters, Washington, D. C.

(*c*) All other files, records, reports, etc., will be forwarded to the Adjutant and Inspector, United States Marine Corps, Headquarters, Washington, D. C., who will transfer to the Paymaster and the Quartermaster such parts as pertain to their departments.

934

CHAPTER 10.

MEDALS AND BADGES.

MEDALS.

1001

Medals of honor, etc.—Any officer or enlisted man may receive a medal of honor, distinguished service medal, navy cross, or life-saving medal. (Arts. 1707, 1709, N. R.)

1002

Brevet commission medals are authorized for issue to officers holding brevet commissions.

1003

(1) *Good conduct medals.*—A marine who has completed four years' continuous service including service on extended enlistments and on reenlistments within four months of discharge, with markings entitlng him to character excellent if discharged, who is distinguished for obedience, sobriety, industry, courage, cleanliness, and proficiency, and who is recommended therefor by his commanding officer, may be awarded a good-conduct medal, or a good-conduct bar if he is in possession of a medal. Awards for periods of less than four years' service will be discontinued, except for men now (Dec. 31, 1921) in the service while in their present enlistment.

(2) An award upon discharge will be made by the officer who issues the discharge certificate; all awards made during an enlistment will be made by the Major General Commandant.

(3) The award of a good-conduct medal upon discharge will be made in the discharge certificate in the following form: "Awarded good-conduct medal (or bar) No. ——, effective on delivery of this certificate." Other awards will be made by good-conduct medal certificate.

(4) *Delivery.*—The medal or bar will be delivered with the discharge or good-conduct medal certificate, or where this is impracticable will be delivered or forwarded as soon as possible thereafter.

(5) *Revocation.*—Good-conduct medals and bars or their advantages can not be taken from men holding them except by sentence of general court-martial.

1004

(1) *Eligibility.*—When the final marking is 4.5 or over (excellent) on expiration of enlistment, a man will ordinarily be eligible for the award of a good-conduct medal or bar.

(2) *Court-martial men not eligible.*—Ordinarily, recommendation for a good-conduct medal should not be made in the case of a man who during his current four-year period of service has been convicted by a court-martial, other than a deck court, and whose sentence has been approved by proper authority; but if a recommendation for a good-conduct medal is made in such a case, the reasons therefor shall be given over the commanding officer's signature.

(3) *Recommendation.*—At posts having a company organization the markings shall be entered personally and signed by the man's company commander, who shall also make recommendations as to good-conduct medals and bars. Commanding officers of such posts shall satisfy themselves that company records are properly kept and that company commanders comply with the provisions of this article. In cases where the commanding officer of a post does not agree with the markings for transfer or discharge or the recommendations in regard to good-conduct medal or bar, his recommendation with reasons therefor shall be given over his signature.

Campaign Insignia.

1005

(1) *Campaign insignia* will be issued to officers and enlisted men who are now or were formerly in the Marine Corps, as follows:

(*a*) *Civil War campaign badge.*—For service in the Marine Corps, Regular or Volunteer Army, or in the militia of the United States, during the Civil War, between April 15, 1861, and April 9, 1865. (N. D. S. O. 82, June 27, 1908.)

(*b*) *Spanish campaign badge.*—For service of not less than 90 days between April 20, 1898, and December 10, 1898, in the Marine Corps. (N. D. S. O. 82, 1908, and N. D. G. O. 81, 1922.)

(*c*) *Philippine campaign badge.*—For service on vessels of the Navy, or on shore in the Philippine Islands, between February 4, 1899, and July 4, 1902, or on shore in the Department of Mindanao cooperating with the Army, between February 4, 1899, and December 31, 1904. (N. D. S. O. 82, 1908.)

(*d*) *China campaign badge.*—For service ashore in China with the Peking Relief Expedition, between May 24, 1900, and May 27, 1901, or with the Legation Guard at Peking, or on vessels of the Navy in China waters as stated in paragraph (d), page 7. (N. D. S. O. 81, June 27, 1908; N. D. S. O. 82, June 27, 1908; act of Mar. 3, 1909; N. D. G. O. 188, Jan. 18, 1916.)

(*e*) *Cuban pacification badge.*—For service in Cuba with the Army of Cuban pacification or in Cuban waters in connection with the Cuban pacification, between September 12, 1906, and April 1, 1909. (N. D. G. O. 35, Aug. 13, 1909; N. D. G. O. 188, Jan. 18, 1916; N. D. G. O. 216, May 23, 1916; N. D. G. O. 381, Mar. 26, 1918.)

(*f*) *Nicaraguan campaign badge.*—For service in Nicaragua under the command of Rear Admiral William H. Southerland or on board the *Annapolis, California, Cleveland, Colorado, Denver, Glacier*, or *Tacoma*, between August 28, 1912, and November 2, 1912, inclusive. (The President's letter to Assistant Secretary of the Navy, Sept. 22, 1913.)

(*g*) *Haiti campaign badge.*—For service in Haiti from July 9, 1915, to December 6, 1915, or any part of such period, or on board those vessels of the Navy listed in N. D. G. O. 305, June 22, 1917, and between the dates set forth in said order for each vessel. (G. O. No. 478, June 17, 1919.)

(*h*) *Haiti campaign badge, 1919–1920.*—For service in Haiti from April 1, 1919, to June 15, 1920, or any part of such period, or on board those vessels of the Navy listed in N. D. G. O. 77, 1921.

(*i*) *Dominican campaign badge.*—For service in Santo Domingo between May 5, 1916, and December 4, 1916, and on board vessels of the Navy listed in N. D. G. O. 76, 1921.

(*j*) *Commemorative expeditionary ribbon.*—For service with the following-named expeditions, for which no campaign badges have been authorized:

Abyssinia, 1903.
China, 1911.
Cuba, 1912, 1917.
Korea, 1903.
Nicaragua, 1909–1910.
Panama, September, 1902; September, 1903; November, 1903; December, 1903; December, 1904; December, 1905; May, 1906; June, 1908; December, 1909.

The ribbon itself represents participation in one of the above-named expeditions. Each additional expedition will be indicated by the placing on the center of the ribbon a metal numeral indicating the total number of expeditions.

(*k*) *Mexican service badge.*—For service on shore at Vera Cruz from April 21 to April 23, 1914, inclusive, or on board any of the vessels named in article A–10 (6), Bureau of Navigation Manual, between the dates designated opposite each ship.

(*l*) *Victory medal, World War.*—For honorable service in the World War, between April 6, 1917, and November 11, 1918. (N. D. G. O.'s No. 482, June 30, 1919; No. 496, Aug. 11, 1919; No. 504, Sept. 27, 1919; No. 508, Oct. 27, 1919; No. 528, Apr. 25, 1920.) M. C. O. No. 50 (1919); N. D. G. O. 83, 1922.

(*m*) *Victory buttons.*—Bronze and silver victory buttons will be issued to officers and enlisted men of the Marine Corps who served honorably therein between the dates of April 6, 1917, and November 11, 1918. M. C. O. 50 (1919).

1006

(1) *French Fourragere.*—The Fifth and Sixth Regiments, United States Marines, having each received three citations in the French Orders of the Army, and the Sixth Machine Gun Battalion, United States Marines, having received two citations in the French Orders of the Army, have been awarded the Fourragere of the colors of the ribbon of the Croix de Guerre (green and red) by the French Ministry of War. This award having been accepted by the War Department on behalf of these organizations, the said Fourragere will become a part of the uniform of the above-mentioned units and will be issued to such officers and enlisted men who are now or may hereafter become members of these organizations. The Fourth Brigade, United States Marines, received one citation in the French Orders of the Army, as an organization; this, however, is the same as that covering the corresponding period for which the Fifth and Sixth Regiments and the Sixth Machine Gun Battalion were cited.

(2) The units above referred to were cited in the French Orders of the Army for their brilliant courage, remarkable ardor and tenacity, valiant advances, and their resolute and energetic activities, as follows:

The Fourth Brigade:
"June 2–13, 1918—Bouresches and Bois de Belleau."
Fifth and Sixth Regiments:
"June 2–13, 1918—Bouresches and Bois de Belleau."
"July 18, 1918—Aisne-Marne (Soissons)"
"October, 1918—Meuse-Argonne (Champagne)."
Sixth Machine Gun Battalion:
"June 2–13, 1918—Bouresches and Bois de Belleau."
"July 18–19, 1918—Aisne-Marne (Soissons)."

(3) In order that an individual may be entitled to wear the said Fourragere at all times, regardless of whether or not he is serving with the unit so decorated, he must have been attached to the organization on at least two occasions covered by the above dates.

MISCELLANEOUS.

1007

Presentation of medals and badges.—The recipient of a medal of honor shall, when practicable, be ordered to Washington, D. C., and the presentation will be made by the President as Commander in Chief, or by such representative as the President may designate. All other medals and badges herein authorized shall, when practicable, be presented by the commanding officer at a parade, and shall be worn on the prescribed occasions.

1008

Wearing medals while undergoing punishment.—The commanding officer may prohibit the wearing of medals by any person undergoing punishment.

1009

Gratuitous issues.—All medals and campaign insignia will be gratuitously issued as an article of uniform to officers and enlisted men who are entitled thereto; also to former officers and enlisted men who have been honorably separated from the naval service. Duplicate medals and campaign insignia will be gratuitously issued to enlisted men in the service, and will be sold at cost price to officers, warrant officers, and men out of the service, upon presentation of satisfactory evidence that the original medal or badge was lost, destroyed, or rendered unfit for use without fault or neglect on the part of persons to whom they were originally furnished, except as otherwise provided in article 1709 (7), N. R.

1010

Officers and enlisted men of the Navy, including former officers and former enlisted men who have been honorably separated from the naval service, and who were attached to marine units on any occasion for which a campaign badge or other insignia is awarded, will be issued the badge or other insignia appropriate to such service upon application to the Major General Commandant.

1011

Badges for marksmanship may be bestowed upon the enlisted men by the Major General Commandant under such rules as may be established with the approval of the Secretary of the Navy.

1012

(1) *Honorable-discharge buttons.*—To each enlisted man of the Marine Corps who may be honorably discharged from the service there shall be issued a Marine Corps honorable-discharge button. This button shall also be issued to each commissioned or warrant officer of the Marine Corps who may be honorably discharged from the service.

(2) To each man or woman, and to each commissioned or warrant officer of the Marine Corps Reserve, who has had three months or more of active service, there shall be issued one of these buttons upon honorable disenrollment from the Reserve.

(3) In cases of men who served one year or more in the Marine Corps and were honorably discharged upon report of medical survey because of disability not incurred in line of duty, and not the result of their own misconduct; because of inaptitude, upon settlement of accounts or by purchase, buttons will

not be awarded, but the requests of the individuals concerned will be given special consideration, and the purchase of honorable discharge buttons may be authorized by the Adjutant and Inspector.

(4) The honorable discharge button adopted by the Marine Corps is intended to be worn only with civilian dress. The face of said button is about nine-sixteenths of an inch in diameter, having in the center the Marine Corps device, with an outer edge of white enamel bearing the words "U. S. Marine Corps—Honorable Discharge" in bronze letters.

(5) The button will be issued gratuitously by the commanding officer of the station at which the man is discharged, with his discharge certificate.

(6) The buttons will be furnished the different stations by the Adjutant and Inspector upon requisition, and will not bear a number.

(7) Ex-enlisted men who have received an honorable discharge from the Marine Corps upon expiration of enlistment will be authorized to purchase buttons from the contractor upon request to the Adjutant and Inspector, who will issue the necessary authority in each case.

CHAPTER 11.

POST FUNDS, ETC.

1101

A post council will be composed of the three officers next in rank to the commanding officer, or of as many as available if less than three. If the commanding officer only is present, he will act. The post council will audit the accounts of the post funds, and will consider all such matters relating to the welfare of the command and to the economy of the post as may be referred to it by the commanding officer, and make such recommendations in regard thereto as it may deem appropriate. The company council will be composed of all commissioned officers present for duty with the company. The composition of the exchange council is prescribed in the exchange regulations (Art. 1207).

1102

Council to audit accounts.—The council pertaining to each organization will audit the accounts and verify the fund pertaining thereto at the end of each month, upon the relief of the custodian and at such other times as it may deem necessary, or as may be ordered by competent authority. It will examine the sources from which and the methods by which the funds have accrued, and will recommend such expenditures as to it appear to be in the best interests of the organization. Each council will meet at the call of its president, a formal order for meeting by the commanding officer not being necessary.

1103

Custodian of funds.—The exchange officer is the custodian of the exchange fund, the company commander of the company fund, the post treasurer of the post funds, and the regimental adjutant of the regimental fund. Custodians of funds shall keep accurate accounts thereof, and no expenditures shall be made which are not solely for the benefit of the organization. The commanding officer will inspect the accounts at least once each month.

1104

The proceedings of each council shall be recorded by the junior member in an appropriate book. Upon the audit of accounts, there will be filed with the record of proceedings, or entered therein, a certificate, signed by the custodian of the fund, of the amount and place of deposit, in a bank or other depository, of the funds or any part thereof, and a statement that the bank and check books have been examined and verified and the cash on hand counted by the council.

1105

The record of proceedings of each council will be submitted to the commanding officer, who will decide on matters of disagreement in the council and will require that the record be kept as prescribed herein. Should the commanding officer disapprove the proceedings or recommendations, and the council after reconsideration adhere to its position, the subsequent action of the commanding officer shall be final, except in cases involving financial responsibility or distribution of profits, which shall be referred to the Major General Commandant, or if the command is a part of a marine brigade serving outside the continental limits of the United States to the brigade commander for final action. The action in each case shall be duly entered in the council record book. The commanding officer will be held responsible for expenditures approved by him which are not in accordance with regulations.

1106

Loss of funds.—In case of loss of any funds, the circumstances shall be carefully investigated and reported by the appropriate council, or by a board of investigation, and recommendations made as to responsibility for the decision of the Major General Commandant or brigade commander.

1108

Forbidden purchases.—The purchase from regimental or company funds of any article which can be obtained on requisition from the quartermaster's department is forbidden, except that, with the approval of the post commander, such articles may be purchased if necessity exists for the immediate use and they are not on hand for issue at the post or if the quartermaster is unable to furnish them after proper requisition has been made.

1109

Forbidden projects.—No project by which money will accrue shall be entered upon under color of military control without specific authority from the Major General Commandant.

1110

Removal of funds from posts.—Under no circumstances shall regimental, post, company, exchange, or other funds be taken away from the post where the organization to which they pertain is stationed, except as may be necessary to pay indebtedness or for a deposit in a bank.

1111

Absence of custodian of funds.—Should the custodian of any of these funds be absent from the post, on leave or otherwise, for any period beyond 3 and less than 10 days, he shall leave the fund with the officer acting in his place, taking memorandum receipt therefor. If the officer be absent for more than 10 days, he shall regularly transfer the fund of which he is custodian to his successor. Company, post, exchange, and other funds shall, if deposited in a bank, be placed under their official designation, as, for example, "Company fund, Marine Barracks, Boston," and not to the credit of the officer who is custodian.

1112

The regimental fund shall consist of the gross amounts received on account of the band and its contributions from the exchange profits, voluntary contributions, sale of articles purchased, or from any other source. The adjutant shall be the treasurer of the fund and shall disburse it under the direction of the regimental commander for the promotion of the efficiency of the band and for such objects as facilitate the transaction of regimental business. A record of all receipts and expenditures and a complete list of all property purchased shall be kept in a regimental fund book.

1113

Extra compensation may be paid from appropriate funds to cooks and mess stewards, when recomended by the appropriate council and approved by the commanding officer. The rates of such extra compensation will be fixed by the appropriate council, with the approval of the commanding officer.

1114

(1) *Post gardens.*—The commanding officer of posts at or near which suitable public lands are available shall set aside for post gardens such ground as may be necessary for the production of vegetables for the command, and shall cause it to be cultivated by the garrison; or, if the commanding officer so elects, he may apportion it among the organizations for cultivation by them.

(2) Surplus products may be sold and the proceeds taken up in the accounts.

1115

Liabilities of company or other funds will be settled as promptly as possible.

1116

Depositories.—When practicable, company, and other funds will be deposited in a Government depository or national bank to the credit of the fund concerned. The check book will be kept as prescribed in the Exchange Regulations.

1117

No expenditure from company fund while receiving Navy ration.—During the periods that an organization receives the Navy ration, no expenditures will be made from the company fund for any article of food except upon special authority of the Major General Commandant, or, if serving outside the continental limits of the United States, of the commanding marine officer.

1118

Upon the abandonment of a post or disbandment of a company or other unit, the commanding officer will direct the closing of the fund accounts, have them audited, and will forward them, with all vouchers, to the Adjutant and Inspector. Funds remaining on hand will be disposed of as the Major General Commandant may direct. The payment of any part of company or other funds to individual men, except for value received, is forbidden.

1119

Cashbook.—The custodians of company, and other funds will keep a cashbook in which the debits and credits must be supported by vouchers. The debit vouchers will consist of dated statements signed by the persons making the payments. Canceled checks or commercial receipts will be accepted as vouchers for credits, but same must be accompanied by itemized bills or by daily memorandum statements to show for what purposes the expenditures were made.

1120

m145

Vouchers, if folded, will be indorsed with the name of the fund, the month, the number or letter, and the amount; if filed flat, the name of the fund, the month, and the number or letter will be plainly entered on the face.

1121

m146

Closing accounts.—The company and other fund accounts will be closed at the end of each month and when the custodian is relieved. The assets and liabilities will be entered in the cashbook immediately after the closed accounts.

1122

m147

Relief of custodian.—When the custodian of a fund is relieved, he will invoice to his successor, and his successor will receipt to him for all funds, accounts, and vouchers turned over, specifying such as are missing, using for the purpose a combination invoice and receipt. This invoice and receipt will be entered or securely pasted in the cashbook, copies being furnished the relieving and relieved officer if desired.

1123

m148

Destruction of records.—Records and accounts of funds may be destroyed, unless sufficient reason exists to the contrary, when four years old, if they have been inspected by an officer of the adjutant and inspector's department and all irregularities have been adjusted.

CHAPTER 12.

EXCHANGE REGULATIONS.

PURPOSE.

1201

(1) *The purpose* of Marine Corps exchanges is primarily to supply the enlisted men at reasonable prices with articles necessary for their health, comfort, and convenience, not supplied by the Government; and, secondarily, through profits, to afford means for recreation and amusement.

(2) The exchange is instituted and maintained for the benefit of the enlisted men, and this principle will be kept in view at all times and under all circumstances.

ESTABLISHMENT.

1202

(1) *For any organization.*—An exchange may be established for any organization of the Marine Corps upon the written application of the commanding officer, approved by the Major General Commandant. Authority to approve the establishment of exchanges for organizations serving outside the continental limits of the United States is delegated to the commanding marine officer of the force with which serving. (See art. 1219 (3).)

(2) *Size of exchange.*—When the establishment of an exchange has been so authorized, the exchange council, with the approval of the commanding officer, will fix the amount of capital necessary, which may be raised by subscription among the enlisted men and officers or may be borrowed from the company funds, or elsewhere. (See art. 1219 (3).) Exchanges will be designated according to the organization to which they pertain, e. g., Post Exchange, Marine Barracks, Norfolk; Second Regiment Exchange; Fourteenth Company Exchange.

(3) *Features.*—No features other than those herein enumerated will be included in an exchange without the approval of the Major General Commandant: (*a*) A store; (*b*) a restaurant; (*c*) amusements, including entertainments, reading, writing, and recreation rooms, library, and games; (*d*) athletics, including teams, prizes, and articles not supplied by the quartermaster's department required for athletic training and for sports; (*e*) barber, tailor, and shoemaker services.

QUARTERS, HEAT, AND LIGHT.

1203

(1) *Buildings.*—Any available set of public buildings, or rooms therein, may be set aside by the commanding officer for the use of the exchange or, when the

financial condition of the exchange will justify such a course, a suitable building or buildings may be erected for the purpose.

(2) If a temporary building is erected for the use of the exchange, or if such a building is constructed wholly or in part by the labor of troops, the use of the necessary teams and such tools and building material that can be spared by the quartermaster's department is authorized.

(3) Repairs and alterations to the exchange building will be made by the quartermaster's department when practicable. When the necessary and authorized repairs and alterations can not be so made, the expense thereof may be borne by the exchange.

(4) *Heat and light.*—The quartermaster's department will provide for the interior and exterior illumination of exchanges, and will also supply the exchanges with such quantities of fuel as may be certified to as necessary by the exchange officer and approved by the commanding officer.

COMMANDING OFFICER.

1204

(1) *Appoints exchange council.*—The commanding officer will appoint the members of the exchange council, the exchange officer, the exchange employees, and the committee of noncommissioned officers. In making these appointments, he should take into consideration the knowledge, experience, and interest of the officers and noncommissioned officers of and in the affairs of the exchange.

(2) *Responsible for expenditures.*—The commanding officer who approves the appropriations of the exchange council will be held responsible for expenditures not made in accordance with regulations.

(3) *General administration.*—The commanding officer will be responsible for the general administration of the affairs of the exchange and will require all regulations pertaining thereto to be properly observed.

EXCHANGE OFFICER.

1205

(1) *Conducts exchange.*—The affairs of the exchange will be conducted by an officer known as the exchange officer, who will be selected and detailed in writing by the commanding officer of the organization to which the exchange pertains.

(2) *Supervises subordinates.*—The exchange officer will carefully supervise the conduct and duties of his subordinates in the exchange. He will frequently and at irregular intervals check their accounts, with a view both to verifying their accuracy and to detecting irregularities.

(3) *Custodian of records, funds, and property.*—The exchange officer will be the custodian of the records, funds, and other property of the exchange and will be responsible for their safe-keeping and preservation. In the case of loss to the exchange, he will be required to show affirmatively that he exercised due care and diligence in the discharge of his duties in all circumstances connected with the loss, failing in which he will be required to reimburse the exchange for such loss.

(4) *Personally takes inventory.*—The exchange officer will take personally such inventories of merchandise and other property as may be prescribed by the exchange council, with the approval of the commanding officer.

(5) *Acting exchange officer.*—In the absence of the exchange officer, the affairs of the exchange will be conducted by an acting exchange officer detailed by the

commanding officer, under such conditions as the commanding officer may prescribe. If the absence is to be for a period of more than 3 days and less than 10 days, the exchange officer will leave the funds with the acting exchange officer, taking a memorandum receipt therefor. If the absence is to be for a period of more than 10 days, the exchange officer will be regularly relieved and the funds and property transferred to his successor.

(6) *Upon relief.*—When an exchange officer is relieved, he will invoice to his successor, and his successor will receipt to him for all accounts and vouchers turned over, specifying such as are missing, using for this purpose a combination invoice and receipt. This invoice and receipt will be filed with the final balance sheet of the officer relieved, copies being furnished the relieving and relieved officers, and the Adjutant and Inspector.

ATTENDANTS.

1206

(1) *Selection.*—The exchange officer will be assisted by a steward and such other attendants as the business may warrant. The attendants will ordinarily be enlisted men detailed from the command, but the employment of retired men or of civilians is authorized where the financial condition of the exchange justifies the expense, in the selection of whom preference will be given to retired and honorably discharged marines, other circumstances being equal.

(2) *Qualifications.*—All attendants of the exchange should be men of excellent record and character. They should be without extravagant tastes and men who are able and satisfied to live within their means.

(3) *Bond.*—The exchange steward will be bonded, and the other employees may be bonded, in an amount to be determined by the exchange council, but at least in sufficient amount to cover their normal activities, with the approval of the commanding officer, the expense thereof to be borne by the exchange.

(4) *The exchange steward,* if an enlisted man, should be a noncommissioned officer having the necessary business qualifications and knowledge of accounts. He should be of unquestioned integrity, have the character necessary to enforce order and discipline in the exchange, and possess the full confidence of the exchange officer in all respects.

(5) *Bookkeeper.*—The employment of a bookkeeper independent of the steward, and with no other exchange duties, is recommended where such a division of labor is practicable. Where no bookkeeper is employed the exchange steward will keep the books of the exchange, under the supervision of the exchange officer.

(6) *Extra compensation.*—The attendants will be paid such extra compensation from the funds of the exchange as may be prescribed by the exchange council, with the approval of the commanding officer.

(7) *Responsibility of steward.*—The exchange steward will be responsible to the exchange officer for the property of the exchange committed to his charge, and will be held pecuniarily responsible for any loss occurring in the exchange due to a failure on his part to exercise due care and diligence in the discharge of his duties. (See arts. 1209 (7), 1212 (9), 1214 (12).)

(8) *Relief of steward.*—When the steward is relieved, an inventory of all merchandise and property will be taken, and his account will be closed. The new steward (and the relieved steward, if practicable) will be present when this is done.

(9) *Personal sales prohibited.*—No attendant will be permitted to sell articles in the exchange on his own account.

(10) Neither the steward nor any other employee of the exchange shall have any personal interest in the purchases, sales, or any advantage of wastage or perquisites of any kind whatever.

EXCHANGE COUNCIL.

1207

(1) *Personnel.*—The exchange council will be a continuous body, and will consist of the exchange officer and of such other officers as the commanding officer may appoint. The minimum membership will be three, unless the number available is less. In case there is no other officer in addition to the exchange officer, the commanding officer will act as a member. In case the commanding officer is alone, he will act as exchange council.

(2) *The exchange council will make recommendations* regarding exceptional purchases, the compensation of attendants and employees, and make such other recommendations as to the management and conduct of the exchange as it may deem appropriate.

(3) *The exchange council will be convened* at any time at the call of its president or by order of the commanding officer.

(4) *Regular meetings.*—On the first day of each month, excluding Sundays and holidays, whenever the exchange officer is relieved and at such other times as may be necessary or advisable, the exchange council will meet and proceed to audit the accounts and take inventories of cash, coupons, bills receivable, and merchandise, and in June and December of all other property. (See art. 1215.)

(5) *The exchange council may delegate* the details of inventory, audit, and investigation of the affairs of the exchange to committees of one or more of its members, excluding the exchange officer, appointed by the president. The reports of these committees will be submitted to the full meeting of the council, and the council in accepting such reports adopts them as its own and becomes responsible for their accuracy.

(6) *The loss of collectible credits will be investigated* and reported on by the exchange council, the report and recommendations of which will be forwarded by the commanding officer, with his recommendations, to the Major General Commandant, or, in the Department of the Pacific, the departmental commander, or, if serving in a marine brigade outside the continental limits of the United States, to the brigade commander for decision as to responsibility.

(7) *The exchange council will verify all entries* in the balance sheets submitted by the exchange officer, and will submit them, over the signatures of all the members, to the commanding officer for his action.

(8) *The proceedings* of the exchange council will be entered in the exchange council book, signed by the president and recorder, and submitted to the commanding officer for his action.

(9) *Minority reports.*—Members of an exchange council have the right to submit minority reports, which will be entered in the council book, signed, and submitted to the commanding officer with the report of the council.

(10) *Disapproval of proceedings.*—Should the commanding officer disapprove the proceedings or recommendations of the exchange council, or any part thereof, he will return the report, with his remarks thereon, for reconsideration and further action. Should the exchange council after reconsideration adhere to its conclusions, and the commanding officer again disapprove, the action of the commanding officer will be final, except as to matters involving financial responsibility and distribution of profits, in which cases the report of the proceedings will be sent by the commanding officer to the Major General Commandant, or, in the Department of the Pacific, the departmental commander, or, if serving with a marine brigade outside the continental limits of the United States, to the brigade commander, whose decision thereon will be final. The final orders in each case will be entered in the exchange council book.

(11) *Responsibility.*—The members of an exchange council will be held pecuniarily responsible for losses to an exchange due to negligence or lack of due care and diligence in the performance of their duties. The following opinion of

the Judge Advocate General of the Army, rendered February 24, 1915, is quoted in this connection:

"POST EXCHANGES: SHORTAGE IN ACCOUNTS; RESPONSIBILITY.

"Upon an examination of the accounts of a certain post exchange the Inspector General's Department found a shortage in the accounts for each month for the period from August 1, 1913, to June 15, 1914, aggregating $655.84. The accounts had not been kept in accordance with the requirements of the post exchange regulations, and it was evident that the loss might readily have been detected by proper auditing of the accounts by the members of the post exchange council, as required by regulations. During the period mentioned the post exchange council took no inventory of the stock, notwithstanding the requirements of the regulations that such inventory be taken by them quarterly or oftener.

"*Held*, That post exchanges, being agencies of the Government, the duties imposed upon officers in the management of their affairs are as binding upon them as any other duty to which they may be assigned under competent military authority; that when the property or funds of an exchange are lost through mismanagement or neglect of such officers, the least that can or should be exacted in the public interests is that they make good the loss; that this principle applies as well to members of an exchange council as to the exchange officer; and that in the instant case it was the duty of the department in the public interests to direct the entry of stoppages against the pay of the several members of the exchange council and of the exchange officer, in equal sums, to cover the shortage."

COMMITTEE OF NONCOMMISSIONED OFFICERS.

1208

(1) A committee of representative noncommissioned officers, appointed by the commanding officer, will be convened quarterly or oftener. This committee will be afforded all proper means for investigating the condition of the exchange and will submit to the exchange council its views and recommendations in respect to the operations of the exchange. The recommendations of this committee will receive due consideration by the council, the action of which thereon will be reviewed by the commanding officer.

(2) The reports of the committee of noncommissioned officers will be signed by its members and entered in the exchange council book immediately after the record of proceedings of the last meeting of the council.

STOCK AND OTHER PROPERTY.

1209

(1) ***Orders for merchandise or supplies*** of any kind for the exchange will be given in writing, signed by the exchange officer, and a duplicate kept on a permanent file. In case a telephonic order is necessary, a memorandum of such order, signed by the exchange officer, will be kept. In cases of emergency, when supplies are required and the exchange officer is not present or available, a written order may be signed by an officer detailed for this purpose by the commanding officer. In no case shall orders, however small, be signed or given by an attendant.

(2) ***Inspection of supplies.***—All merchandise and other property received for the exchange will be inspected by the exchange officer, or in case he is not available by an officer designated by the commanding officer, who will personally satisfy himself as to the quantity and quality of each article received, and

certify to the facts on the face of the invoice. If no invoice is received a certificate will be prepared and signed by the inspecting officer. A rubber stamp of the following form should be provided:

Invoice No.________________
Invoice received ________________
Goods received________________
Checked in ________________
Entered S. R.________________
Entered steward________________
Bookkeeper________________
Paid by check________________
Voucher No.________________

(3) *Purchases limited.*—Supplies will be purchased only in sufficient quantities to meet the needs of the exchange for the immediate future. The amount carried will be governed by the proximity of adequate markets and the facilities for delivery to the exchange.

(4) Merchandise in exceptional quantities will not be procured except upon the recommendation of the exchange council, with the approval of the commanding officer. (See art. 1207 (2) and (8).)

(5) *The quartermaster's department is authorized to sell* for cash to exchanges at cost such articles of clothing, rations (including ice), forage, furniture, and fixtures as may be needed and can be spared from the stock on hand.

(6) *Storeroom.*—In exchanges where such a system is practicable, a storeroom should be provided, which will be placed in charge of an attendant as storeroom keeper.

(7) When this is done, the storeroom keeper will receive all stores of whatever kind that comes into the possession of the exchange, and issue all stores to the several departments on written requisition in triplicate. These requisitions will be O. K'd by the steward, and when filled the original will be filed in the storeroom, the duplicate with the department receiving the goods and the triplicate with the person who keeps the accounts. (See art. 1214 (13, 16) and 1217 (6).)

SERVICES.

1210

(1) *The barber, shoemaker, and tailor services* to enlisted men at all shore stations of the corps will be conducted by the exchange, and such enlisted men or other persons who may perform these services within such commands will be employed by the exchange upon such terms as may be prescribed by the exchange council, with the approval of the commanding officer.

(2) The barber, shoemaker, and tailor may be paid either a salary or a percentage of the amount of their services, as may be determined by the exchange council, with the approval of the commanding officer.

MANAGEMENT.

1211

(1) *All obligations will be paid as soon as practicable,* to the end that the liabilities of the exchange may be at a minimum at all times, and the advantages of discounts obtained.

(2) *Price lists* will be conspicuously posted in the various sections of the exchange.

(3) *A copy of the latest balance sheet,* showing the exchange council's and commanding officer's action thereon, will be kept posted in a conspicuous place in one of the exchange rooms.

(4) ***Regulations.***—A corrected copy of these regulations will be kept hung up at all times in an accessible place in one of the exchange rooms.

(5) ***Rules of order*** will be prescribed by the exchange officer, with the approval of the commanding officer, and a copy thereof posted in each of the exchange rooms.

(6) ***Gambling forbidden.***—Gambling or playing any game for money, or anything of value, or raffling, is forbidden in an exchange.

(7) ***Sale of intoxicants forbidden.***—The sale of or dealing in beer, wines, or any intoxicating liquor by any person in any exchange is prohibited.

(8) ***The use of penalty envelopes*** will be limited strictly to the proper correspondence of the exchange, and will not be used in soliciting custom nor in the delivery of goods. Return penalty envelopes are not authorized.

(9) ***Civilians*** not employed at a post will not be permitted to enter an exchange without the authority of the commanding officer.

(10) ***The presence of an attendant*** in the exchange will be required all day, and at night when at all practicable.

CASH.

1212

(1) ***The exchange officer will attend to all cash transactions*** in person, except routine cash sales and routine settlement of personal accounts receivable, and he will allow no employee to have access to the cash after it is turned in to him. A reasonable sum, for which monthly receipt will be taken, may be placed or left in the hands of the steward for the purpose of making change.

(2) ***Removal of funds from station.***—Under no circumstances will the exchange officer remove the exchange funds from the station of the organization or post to which they pertain, except as may be necessary for the payment of obligations, for deposit, or for the purpose of obtaining necessary change.

(3) ***Depository.***—When not impracticable, exchange funds, except such amounts as may be reasonably necessary for routine transactions, will be deposited in a Government depository, or if one is not available, in a national bank.

(4) ***Government checks may be cashed*** by the exchange officers for enlisted men, and will be cashed for marines discharged, if practicable, whenever such checks can not be cashed otherwise without expense or delay. No charge will be made for this accommodation.

(5) ***Private checks*** may be cashed by the exchange officer for officers in amounts fixed by the exchange council, with the approval of the commanding officer, but not in excess of $25.00 in one day for any officer.

(6) ***Payments of bills by check.***—The exchange officer will not permit the attendants to pay the obligations or bills of the exchange and will, whenever it is possible, make payments of such obligations or bills by check.

(7) ***A cash register*** should be provided for each department of the exchange, if practicable. All cash received will be placed in the cash drawer and the amount of the sale rung up. Where practicable the indicator of the cash register will be kept locked and the keys kept in the personal possession of the exchange officer.

(8) The exchange officer will compare daily the sales as shown by the cash register with the sales as shown by the steward's daily report.

(9) ***Cash turned in by steward.***—The commanding officer shall designate an hour at which the steward will turn in daily to the exchange officer the cash received during the day prior to that hour. On days when unusual amounts are received, cash will be turned in to the exchange officer as many times as may be necessary, in order to prevent a considerable accumulation of cash in the hands of the attendants. At no time shall the total cash in the custody of the steward be allowed to exceed one-fourth the amount for which the steward is bonded.

(10) *Cash will not be left in the cash register or drawer overnight.*—In exchanges where the steward is provided with a combination lock safe of such size and condition as to be reasonably secure against theft, such cash as may be in the hands of the steward for change, as provided in paragraph (1), together with cash that is received between the hour mentioned in paragraph (9) and the close of business for the night, will be placed in the steward's safe, provided the total of such cash does not in any case exceed one-fourth the amount for which the steward is bonded. When cash is thus left in the safe of the steward overnight, all cash receipts so left will be turned in to the exchange officer as soon as practicable on the following morning.

(11) *In the absence of the exchange officer* at any of the times above mentioned requiring the cash to be turned in, the steward will turn in the cash received to the officer of the day, or to such other officer as may be designated by the commanding officer, who will give receipt for same and be responsible for its delivery to the exchange officer.

CREDIT.

1213

(1) *No credit ordinarily.*—All Marine Corps exchanges will be conducted on a strictly cash basis, extending no credit whatsoever for merchandise or services, except as provided in paragraph (2).

(2) *When credit is extended.*—Exchanges established for troops serving in the field beyond the continental limits of the United States may, if it be a military necessity, extend credit, not exceeding one-half of one month's pay, to enlisted men, not more than one-half of such credit to be extended during the first 15 days of the month, and a reasonable credit to officers. At other posts, enlisted men not in good standing, prisoners not in a pay status, applicants awaiting enlistment at recruit depots, men joining by reenlistment after having been separated from the service for a period of more than one month, may be extended credit by the exchange, on the written order of the commanding officer, for such toilet articles, barber, tailor, and shoemaker services as may be actually necessary for health and comfort. Such credit will be kept within a reasonable limit, the maximum to be fixed by the exchange council, with the approval of the commanding officer. Credit extended applicants at recruit depots will, if the enlistment is accomplished, be checked on the first pay roll on which the recruit's account appears. If the enlistment is not accomplished, the amount may be dropped as a loss.

(3) *Services furnished prisoners gratis.*—Where the services of the tailor, barber, and shoemaker are performed by enlisted men, employees of the exchange, for enlisted men held in custody as prisoners not in a pay status who are awaiting a trial or final disposition of their cases, neither the exchange nor its employees will make any charge for such services nor for the actual cost of supplies or materials furnished in connection with such services.

(4) *Collections of indebtedness* as above authorized of enlisted men to exchanges will, where practicable, be made through the paymaster's department by checkages on pay rolls in accordance with instructions issued by the Paymaster. The amount of indebtedness to an exchange shall not be collected until all stoppages for indebtedness to the United States have been made and all forfeitures by sentences of court-martial, if any, have been satisfied.

(5) *Procedure for extending credit.*—In cases where paragraph (2) is applicable, the following procedure will be followed:

(*a*) At the beginning of a month, each officer rendering a pay roll will submit a credit roll (Form C) to the exchange officer upon which will appear the names of the enlisted men of the organization for which the pay roll is rendered who desire credit at the exchange, with the amount of credit approved for each for the month. Supplementary credit rolls will be submitted for men who join during the month subsequent to the submission of the credit roll.

Where collections are made through the paymaster's department, the credit roll will be submitted in quadruplicate, one copy to be retained by the exchange officer as his authority for extending the approved credits, one copy to be returned immediately with a certificate of the receipt of copies, and at the close of the credit period of the month, which will be prior to the closing of the pay roll, two copies to be returned with the amount of credit extended to each man during the month entered opposite his name in the proper column. Of the two copies received by the officer rendering the pay roll at the end of the credit period one will be retained by him as his authority for making the proper checkages on the pay roll, and the other returned by him to the exchange officer with a certificate of receipt of copy. Where collections are made in cash, the credit roll will be submitted in duplicate, one copy of which will be retained by the exchange officer as his authority for extending the approved credits and one copy returned with a certificate of receipt of copy.

(*b*) Officers commanding organizations for which pay rolls are rendered will submit to the exchange officer, in time to afford the latter opportunity to take the required action, a written notice (Form A), in duplicate, of change in the credit status of any enlisted man of the command for which the pay roll is rendered, due to checkage, prospective checkage, confinement, trial by court-martial, transfer, discharge, death, retirement, desertion, or to any other cause within their knowledge. The exchange officer will retain one copy of this notice and will return one copy with certificate of receipt of copy.

(*c*) The exchange officer will submit to commanding officers of organizations for which pay rolls are rendered written notice (Form B), in duplicate, of indebtedness to the exchange due to credit extended to enlisted men of the respective commands in all cases where checkage is required and which are not provided for on the credit rolls. One copy of this notice will be retained by the commanding officer of the organization and one copy returned to the exchange officer with a certificate of receipt.

(*d*) Officers commanding organizations for which pay rolls are rendered will be held liable to an exchange for loss due to extension of credit approved by them in excess of that authorized by regulations, or to failure to give due notice as required in paragraph (*b*), or to failure to make proper checkages on the pay rolls, or to enter indebtedness in service-record books upon due notice. The exchange officer will be held liable to the exchange for loss to the exchange due to the extension of credit in excess of that approved on credit rolls, or to the extension of credit of more than one-half of the authorized credit on or before the 15th of the month, or to failure to enter the proper amount of credit extended on credit rolls, or to failure to give due notice of credit extended as required in paragraph (*c*), or to make the proper collection or claim upon due notice, as required in the instructions of the paymaster's department.

(*e*) Where collections are made in cash a collection roll (Form D) will be prepared for use on pay day, showing the amount of indebtedness of each man. Opposite the column of amounts due will be a column for each day on which collection is to be made, and as payments are made the amounts will be entered opposite the proper name in the paid column of the proper date. The totals of these columns will show the total amount of collections each day.

(*f*) Credit and collection rolls and credit notices will form part of the permanent records of the exchange.

(*g*) Sales slips (Form F) will be used in all exchanges in the extension of credit to enlisted men not in good standing.

(*h*) Sales slips will be initialed by the salesman and signed by the purchaser. In the cases of enlisted men not in good standing, they will show the approval of the commanding officer. (See art 1213 (2).) They will be retained by the exchange officer as notes until the indebtedness has been satisfied, when they will be plainly stamped "Paid" and returned to the men if practicable, or, if not practicable, destroyed, except that when men are transferred such slips will be retained six months before being returned or destroyed.

(*i*) Sales slips received will be turned in by the steward to the exchange officer daily.

(*j*) Coupons will be used, when practicable, for the extension of credit to enlisted men in good standing. (See art. 1220.)

ACCOUNTS AND RECORDS.

1214

(1) *How kept.*—The accounts of an exchange will be kept in such form and in such detail as will enable the exchange officer, the exchange council, the commanding officer, or an inspecting officer to obtain a full history of the transactions of the exchange, and to ascertain the condition of the affairs of the exchange at any time.

(2) It should be kept in mind that transactions which are not recorded are not available for scrutiny in the absence of those actually effecting the transactions, and that therefore all essential matters should be committed to the records.

(3) *Removal of records from station.*—The records of an exchange will not be removed from the station of the exchange except on the authority of the commanding officer or the Major General Commandant. Upon the permanent closing of the exchange they will be forwarded to the Adjutant and Inspector.

(4) *The double-entry system of bookkeeping* will be used in all exchanges, and the following books and accounts will be kept, except as noted:

(5) *Blotter,* in which the exchange officer will keep in his own handwriting a rough record of all cash receipts and expenditures. The blotter may be omitted if the exchange officer keeps the cashbook-journal in his own handwriting.

(6) *Cashbook-journal* (Form G), in which will be kept a smooth account of all transactions. Entries will be made in separate columns, appropriately headed, according to the classification into which it is desired to divide the receipts and expenditures. The number of the check or voucher supporting each expenditure will be entered in the appropriate column.

(7) *Ledger,* in which summaries of the exchange's debit and credit transactions will be entered. The ledger will be divided into two sections—the general and the petty ledger. The following accounts will be kept in the general ledger:

(*a*) *Amusements.*—This is purely a loss account and will show on the debit side the total of all donations—either cash, merchandise, or anything of value—to amusements. The credit side will, as is done under the "Expense" account, be balanced to "Loss and gain." The total of the debit entries is the total as taken from the "Amusement" column on the debit side of the cashbook journal.

(*b*) *Amusement rooms.*—This account must not be confused with the "Amusement account." This account covers the activities of the billiard room and bowling alleys and other forms of amusement from which revenue is derived. On the debit side will appear the total expenses involved in conducting this activity. On the credit side will appear the total of the business done (cash and credit), and the balance (red ink) debited or credited to the "Loss and gain" account.

(*c*) *Bills receivable.*—(Due from enlisted men.) The debit side of the account will contain the balance due under the head at the beginning of the period; the total of all sales on credit to enlisted men during the period. (This last entry is the total of the column headed "Bills receivable" on the debit side of the cashbook journal.) The credit side will show the total of all cash receipts under this head during the period. This is total of column headed "Bills receivable" on credit side of cashbook journal. The credit side will also show (in red) the balance due under this account, which balance must agree with the schedule of bills receivable filed with the balance sheet. (See art. 1215 (20).)

(*d*) *Cash.*—This account will contain on the debit side the cash balance on hand at the beginning of the month or other accounting period; also the sum of all items of cash received during the period (this is total of cash column on debit side of cashbook-journal). The credit side will contain the sum of all items of cash actually paid out during the period (this is total of cash column on credit side of cashbook-journal); also (in red ink) the amount of cash remaining on hand at the end of the accounting period.

(*e*) ***Coupons.***—On the debit side of this account will appear the amount of coupons on hand at the beginning of the period; also the amount of coupons received and redeemed. The credit side will show the amount of coupons issued during the period. The balance (red ink) will show the amount of coupons outstanding at the end of the period.

(*f*) ***Exchange account.***—This account will show the value of the exchange. On the debit side will be entered the net loss, if any, and (in red ink) the present worth. On the credit side will be entered the value at the beginning of the period and the net gain, if any.

(*g*) ***Expense.***—This account will contain on the debit side the total of all items of expense incurred during the period. The following items properly belong to this account: Compensation of attendants, bonding employees, insurance, books and stationery for use in exchange, ice, and transportation charges *only* when the transportation involved is on an item belonging to this account. The debit entry will be the *total* of the expense column debit side of cashbook-journal. The credit side of expense account will contain a red-ink entry "Loss and gain," and the same amount that is shown on the debit side. This is because there is no way whereby any income can be had under expense, and it is always a loss.

(*h*) ***Interest and discount.***—This account being normally the opposite in character from the expense account, showing usually a gain without a corresponding loss, shows on the credit side the total of all items of interest received on deposits, and all items of discount actually taken on paid bills. This entry is the total of interest and discount column on credit side of cashbook-journal. The debit side shows in *red* the same amount to "Loss and gain."

(*i*) ***Loss and gain.***—The debit side of this account will contain the total of the loss and gain column of the cashbook journal, the items of that column showing the character of the losses sustained in conducting the business. The credit side will contain the total of the loss and gain column on the credit side of the cashbook journal, this column showing the character of the gains during the period covered. In addition, the profit in merchandise or property is entered on the credit side, or in case of a loss therein as shown by the inventory such loss is entered on the debit side. The balance (in red) of this account, showing the net loss or net gain, is entered on the side necessary to balance. If this balance is on the credit side it is a net loss, if on the debit side it is a net gain. The net loss or net gain is carried to the "Exchange account (in red ink)."

(*j*) ***Merchandise.***—This account will contain on the debit side the total of the merchandise (cost value) on hand as shown by the inventory at the beginning of the month or accounting period; also the total of all amounts, including any transportation charge on merchandise (this is total of merchandise column on debit side of cashbook-journal). The debit side also will contain (in red ink) the profit on merchandise during the period. The credit side will contain the amount of merchandise sold (selling value) during the period, and the cost value of all merchandise expended, returned to dealer, or transferred to other accounts (see art. 1217 (6)) (this is total of merchandise column on credit side of cashbook-journal); also the inventory (cost value) of merchandise on hand at end of period (in red ink).

(*k*) ***Property.***—This account will show transactions in all items concerning property of the exchange which is not carried for sale. The account will show on the debit side the value of all property on hand at the beginning of the accounting period; also the value of all property received during the period; the amount (if any) of any increase in value during the period, such as a sale at a higher amount than the paper value. On the credit side will appear the value of all property expended; the amount realized from the sale of property; the amount of depreciation during the period; and (in red ink) the value of the property on hand at the end of the period.

(*l*) ***Barber shop.***—This is an *impersonal* account and must not be confused with the personal account of the employee who performs the services therein. On the debit side will appear the total expenses involved in conducting this activity. On the credit side will appear the total of the business done (cash and credit), and the balance (red ink) debited or credited to the "Loss and gain" account.

(*m*) ***Shoemaker.***—This is an *impersonal* account and must not be confused with the personal account of the employee who performs the services therein. On the debit side will appear the total expenses involved in conducting this activity. On the credit side will appear the total of the business done (cash and credit), and the balance (red ink) debited or credited to the "Loss and gain" account.

(*n*) ***Tailor shop.***—This is an *impersonal* account and must not be confused with the personal account of the employee who performs the services therein. On the debit side will appear the total expenses involved in conducting this activity. On the credit side will appear the total of the business done (cash and credit), and the balance (red ink) debited or credited to the "Loss and gain" account.

(*o*) ***Accounts payable.***—On the debit side will be entered the total of all payments made by the exchange to dealers on any merchandise received from the dealer (this is the total of accounts payable column on debit side of cashbook journal). The credit side will show the balance (if any) owed by the exchange to dealers or firms at the beginning of the period, also the total of purchases made by the exchange and total of all other bills incurred by the exchange, whether paid during the period or not (this is the total of "Accounts payable" column on credit side of cashbook journal). The balance of this account (red ink) should agree with total of "Schedule of accounts payable" filed with balance sheet. (See art. 1215 (22).)

(*p*) ***Accounts receivable.***—The debit side of this account will show the balance due the exchange under this head at the beginning of the period; the total of all credit sales to officers (this total is the total of accounts receivable column on debit side of cashbook journal). The credit side will show the total of all payments on account made by officers during the period and (in red) the balance due the exchange under this head.

(8) ***The petty ledger*** will contain accounts with every firm or individual with whom the exchange has any dealings. This includes all employees and attendants. The debit side will contain, first, any amount that may be owed to the exchange by such firm or any individual at the beginning of the period. (The total of such entries should equal the total of the "Schedule of accounts receivable" filed with the balance sheet for the prior period.) The debit side will also show all payments made to the firm or individual concerned as shown by the cashbook-journal. The credit side will show first the amount owed *to* the firm or individual at the beginning of the period, for merchandise, services, or anything else. (The total of these entries must agree with the total of the "Schedule of accounts payable" filed with the balance sheet for the prior period.) The credit side will also show all transactions wherein the firm or individual concerned has furnished or supplied the exchange with merchandise, services, or anything else of value. The balance of each account (in red ink) will show, if such balance is entered on the debit side, the amount the exchange owes that person at the end of the period; if the balance appears on the credit side, the amount the person or firm owes the exchange. The total of all the above balances that appear on the debit side must agree with the "Schedule of accounts payable" filed with the balance sheet. The total of all balances appearing on the credit side must agree with the "Schedule of accounts receivable" filed with the balance sheet.

(9) *Invoices.*—There will be entered on the face of invoices the inspector's certificate (art. 1209(2)), the selling price of each group of articles, the total selling value, and the account or accounts, with the amount of each, to which debited. Invoices will be numbered serially by months, and the number of each invoice will be entered opposite the debit in the ledger accounts.

(10) *Vouchers.*—Vouchers will be required for all cash expenditures. Paid checks will be accepted as vouchers for expenditures where the accounts show the purpose for which the expenditures were made. When bills are not paid by check, commercial receipts or receipted bills will be required. Vouchers will be plainly numbered by months, and the voucher number entered opposite each expenditure entered in the cashbook-journal.

(11) Separate permanent flat files will be kept for invoices, vouchers (except paid checks), inventories, balance sheets, steward's daily reports, credit rolls and notices, and collection rolls, on which these records will be filed in order.

(12) *Steward's daily reports.*—The exchange officer will require the steward to submit to him daily (Form H) a report of each day's business, showing the sales and services for cash, coupons, and on credit, the cash on hand last report, cash since received, cash turned in and remaining on hand, merchandise on hand last report, since received, sold, returned, and expended or transferred, and merchandise remaining on hand, and coupons on hand last report, since received, since issued, and books remaining on hand. These reports will be prepared and submitted, whether or not any business was transacted, examined, and, if found correct, certified by the exchange officer.

(13) *Steward's account.*—The exchange officer will keep in his own handwriting a memorandum account with the steward in which all entries will be made in the presence of the steward charging him with the selling value of the merchandise turned over to him for sale, and crediting him with the selling value of all goods sold or properly expended. Proper debit or credit will be made when selling prices are changed. The balance of this account will show the selling value of the merchandise which the steward should have on hand, which should check with the inventory selling price.

(14) *The bank balance* will be kept in the bank check book. Deposits and amounts drawn by check will be entered on the stubs and the balance carried forward on each page. On the last day of each month, and at such other times as a statement is obtained from the bank, the number and amount of each outstanding check, the total of the outstanding checks, and the difference between the balance as shown by the bank statement and the total of the outstanding checks will be entered on the back of the last check stub for the period covered by the bank statement. This difference should equal the balance as shown on the check stub.

(15) *Checks will be numbered serially,* and stubs will show date, number, name of payee, and amount. Canceled checks will be attached to stubs and the word "Canceled" written across the face of the check and of the stub. Paid checks returned from the bank will be immediately pasted or otherwise securely attached to their stubs in such a manner as to not cover any entries on the stub.

(16) *Stock book.*—The storeroom keeper will enter in a stock book or card file, on the debit side, the selling value of all stores received, and on the credit side the selling value of all stores issued. This account may be kept on the basis of units of items issued without the value appearing therein.

(17) All the above-mentioned records and accounts must be kept, except as noted, and such other books and accounts as may be necessary will also be kept. Any deviation from these regulations in regard to accounts will require the approval of the Major General Commandant.

(18) *Destruction of records.*—Records and accounts of the exchange may be destroyed, unless sufficient reasons exist to the contrary, when four years old, if they have been inspected by an officer of the adjutant and inspector's department and all irregularities have been adjusted.

INVENTORY AND AUDIT.

1215

(1) ***The value of an inventory*** depends on the thoroughness and accuracy with which it is taken, and errors of omission or commission may result in injustice either to the exchange personnel or to the command, or to both. Therefore the greatest care and exactness will be exercised in taking inventories.

(2) ***In taking inventories*** the services of officers or of competent enlisted men, except the exchange officer and attendants, may be utilized. The counting, recording, and extensions must be performed under the direct and personal supervision and scrutiny of the person or persons responsible for the inventory. A sufficient number of extensions, chosen at random, should be personally checked in order to establish a strong probability that they are being correctly made. A sufficient number of unit cost prices should likewise be personally verified, in order to insure that a proper value is being given to the stock. Inventories (Form I) will be signed by the members of the inventory committee.

(3) ***In auditing the accounts*** of the exchange the services of qualified officers or noncommissioned officers or privates may be utilized, excluding the exchange officer and attendants. All work connected with the auditing must be done under the direct and personal supervision and scrutiny of the members of the auditing committee.

(4) ***The cash,*** including that in the hands of the steward for change, will be counted at the beginning of the audit. Statements will be obtained from the bank and the balance on deposit, as shown by the last stub, will be verified; entries on stubs showing deposits will be compared as to date and amount with such entries on statement and in pass book.

(5) ***Colored pencils.***—All persons conducting the audit will be provided with blue pencils. The exchange officer and the exchange employees will use ordinary black pencils for checking items. Green pencils will be reserved for use by officers of the adjutant and inspector's department.

(6) ***In auditing the accounts*** the records should be distributed among the auditors and assistants as follows:

(*a*) Cashbook journal.
(*b*) Exchange officer's blotter.
(*c*) Steward's daily reports.
(*d*) Other cash receipt vouchers.
(*e*) Cash expenditure vouchers and check book.
(*f*) File of orders for merchandise or property.
(*g*) Invoices for merchandise and property.
(*h*) Steward's account. (See art. 1214 (13).)
(*i*) Ledger.

This distribution may be varied according to the number of auditors and assistants available or other circumstances in each case.

(7) The person having the cashbook-journal will verify the entries therein as to date, amount, and classification as the items are called to him by persons having the other records.

(8) The person having the exchange officer's cash blotter will read from it the entries therein by date, item, and amount.

(9) The persons having the steward's daily report and other cash receipt vouchers will read from these the data supporting the entries in the cashbook-journal under the following heads: Cash (debit); accounts receivable (debit and credit); merchandise (credit); bills receivable (debit and credit); barber, tailor, and shoemaker (credit); coupons (debit and credit); property (credit); amusement rooms (credit); and sometimes accounts payable (credit).

(10) The persons having cash expenditure vouchers and check book will read therefrom the data supporting the entries in the cashbook-journal under the following heads: Cash (credit); accounts payable (debit); barber, tailor, and

shoemaker (debit); expense (debit); amusements (debit); interest and discount (credit); property (debit) and merchandise (debit).

(11) The person having the file of orders for merchandise or property will, with the person having the invoice file, compare the orders with the invoices, noting carefully that the regulations have been observed. (See art. 1209 (1) and (2).)

(12) The person having the invoice file will then read therefrom the data supporting the entries in the cashbook-journal under the following heads: Merchandise (debit); property (debit); accounts payable (credit); interest and discount (credit), reference being also necessary to the cash expenditure vouchers for items under interest and discount.

(13) Items in the cashbook-journal showing loss and gain, being supported by various records depending on the character and cause of the loss or gain, must be determined and verified by examination of the circumstances in each case.

(14) As each entry in any record is verified a check mark (√) will be made opposite.

(15) All additions on cashbook-journal will be verified and a preliminary test balance made of all entries therein.

(16) The totals of all columns in cashbook-journal are next verified as having been posted to the proper account in the general ledger. Postings to the individual accounts in the petty ledger (accounts receivable and accounts payable) need not be verified in detail unless some special reason therefor develops. However, a test will be made by totaling all credit and debit entries therein and comparing the difference between these totals with the general ledger balances under these heads.

(17) After checking each book, file, and account they will be gone over by a member of the auditing committee to ascertain whether any items are not checked, and if such items are found they will be looked up.

(18) The additions in each account will be verified, and the balances as brought forward and as carried to other accounts will be checked and verified. The balances will be compared with those entered in the balance sheet. The balance in the steward's account (exchange officer's) will be compared with the inventory, selling price.

(19) Auditing committees will take especial care to see that adding machines are not improperly manipulated, it being possible on some machines to obtain false totals by manipulation of the eliminating key.

(20) Notes will be counted, compared with the credit rolls and statement of indebtedness on balance sheet, and the total amount compared with the amount called for by the bills-receivable account in the ledger. If any reasons therefor should appear, the validity of the notes will be looked into and verified in such manner as is most feasible.

(21) The steward's coupons on hand will be counted to see if they agree with the amount as shown by the daily report, and as shown by the exchange officer's account with him. Coupon books will be examined to ascertain if any coupons have been removed.

(22) The amounts shown as assets and liabilities will be carefully checked. The sums shown as accounts receivable and payable will be carefully checked with the total amounts of personal accounts as shown by the ledger as due to and from the exchange.

(23) The other items of the balance sheet will be carefully checked with the accounts in the ledger as they appear at the time of the audit.

(24) Any and all discrepancies or irregularities will be looked into with care and report made as to responsibility.

(25) The council will carefully examine into the schedule of accounts payable and accounts receivable filed with the last prior balance sheet and ascertain whether or not any remain unpaid and, if any such are found, the reason therefor investigated and recorded.

DISTRIBUTION OF PROFITS.

1216

(1) *Determination of profits subject to distribution.*—Profits will be distributed monthly. Profits subject to distribution will be the amount of cash remaining on hand after deducting the following sums:

(*a*) A sum sufficient to pay all liabilities of the exchange.

(*b*) A sum which, added to 90 per cent of the total bills and accounts receivable at the end of the month, will be sufficient to pay all anticipated ordinary expenses (including operating expenses, purchases of stock, and other property) for the ensuing month.

(*c*) With the approval of the Major General Commandant, or, if serving in a marine brigade outside of the United States, of the brigade commander, a sum which may be required for some specific object for the betterment of the command additional to the anticipated ordinary expenses.

(*d*) The amount of profits subject to distribution is determined as follows:

Total cash on hand May 31, 1918	$2,450.00
Total liabilities May 31, 1918	35.60
Cash on hand over liabilities	2,414.40
90 per cent bills and accounts receivable May 31, 1918	385.60
	2,800.00
Less anticipated ordinary expenses for June, 1918	2,000.00
Amount of cash profits subject to distribution on May 31, 1918 (payable in June)	800.00

(*e*) There are no profits subject to distribution when the total liabilities exceed the amount of cash on hand.

(2) Profits subject to distribution will be credited to the proper accounts for the recreation and amusement of the enlisted men in such proportion as the exchange council, with the approval of the commanding officer, may direct.

(3) A division of profits, to be determined as required by paragraph (1), will also be made when 25 or more men are transferred to expeditionary service. For each enlisted man so transferred a proportionate share of the profits as of the date of transfer will be sent as soon as possible to the senior marine officer of the expedition, who will distribute the amount so received pro rata to the company funds of the organizations to which the men transferred are assigned.

(4) When an entire organization is transferred to expeditionary duty, the proportionate share of the exchange profits will be paid into the organization's funds.

(5) *Payment to individual men prohibited.*—The payment of any part of exchange profits to individual men is prohibited except in the nature of prizes as outlined in art. 1202(3).

(6) Recommendations for deviation from the methods provided for the distribution of profits may be submitted for approval to the Major General Commandant, or, when serving in a marine brigade outside the continental limits of the United States, to the brigade commander.

REPORTS.

1217

(1) The exchange officer will close his books at the end of each month's business, when relieved, and at such other times as may be directed by proper authority, and will submit, in triplicate, to the exchange council a balance sheet

(Form J) to include that date, giving in detail all of the information required thereon.

(2) After auditing the accounts the exchange council will submit the balance sheets, with their report, to the commanding officer for his action.

(3) After the commanding officer has taken final action on the report, he will forward the original of the balance sheet to the Adjutant and Inspector and return the two copies to the exchange officer.

(4) The exchange officer will place one copy of the balance sheet on a permanent file, and will post the other copy in a prominent place in one of the exchange rooms for the information of the enlisted men of the command.

(5) A schedule of personal bills receivable, accounts receivable and payable, showing dates each are due, will be forwarded with the original of the balance sheet and a copy attached to the file copy with the records of the exchange.

(6) A statement of all transactons affecting the credit side of the Merchandise Account (other than cash sales) will be shown on the back of, or attached to balance sheets. These transactions should be shown under appropriate headings and should include the following:

(*a*) Merchandise returned to dealer.
(*b*) Merchandise donated to Amusements.
(*c*) Merchandise transferred to Property.
(*d*) Merchandise transferred to Expense.
(*e*) Marchandise transferred to any other account.

Final Disposition of Business.

1218

When it becomes known that an exchange is to be discontinued the exchange stock will be reduced to the lowest extent possible and, so far as may be, converted into cash. Prior to the closing of the exchange the property will be converted into cash or otherwise disposed of, as directed by the Major General Commandant. Such disposition of the cash will be made as may be directed by the Major General Commandant.

Headquarters Marine Corps.

1219

(1) *The Marine Corps exchange board* will be composed of the Headquarters exchange officer and of such other officers as the Major General Commandant may designate. The duties of this board will be to consider and make recommendations upon such subjects as pertain to the affairs of the Marine Corps exchanges and to audit accounts of the Headquarters exchange officer. This board will meet at the call of its president or upon the order of the Major General Commandant, and a record of its proceedings will be kept by the Headquarters exchange officer for submission to the Major General Commandant for his action.

(2) *The Headquarters exchange officer* will be designated by the Major General Commandant. He will have general supervision of correspondence pertaining to matters connected with exchanges under the supervision and orders of the Adjutant and Inspector.

(3) It will be his duty to take charge of and administer the Marine Corps fund. This fund will consist of all moneys received from exchanges of the Marine Corps and of such funds as may be received from other sources. The Marine Corps fund will be disbursed for the general benefit of the Marine Corps, as may be deemed appropriate by the Marine Corps exchange board and

approved by the Major General Commandant. The fund is available for furnishing capital for the establishment of news exchanges and similar purposes.

COUPON BOOKS.

1220

Coupon books will be serially numbered and be countersigned before issue by the exchange officer. The name of the man to whom issued will be entered on the cover, and the coupons will not be honored except when presented by the man to whom issued. The books will be carefully accounted for, both by the exchange officer and the steward. (See art. 1214(7n), 1214(12) and Form "E.") Other forms will be substantially as indicated below, and will, in any event, contain all information and data required therein.

FORMS.

1221.

FORM A.

NOTICE OF CHANGE OF CREDIT STATUS.

27th Company,
Marine Barracks, Washington,
August 3, 1915.

MEMORANDUM FOR EXCHANGE OFFICER.

Private Henry Clayton will be transferred at 3:00 P. M. to-day to navy yard, Washington, D. C.

JOHN DOE, Commanding (2 copies).

Copy received,

RICHARD POE, Exchange Officer (1 copy).

FORM B.

NOTICE OF CREDIT EXTENDED.

Post Exchange,
Marine Barracks, Washington,
August 3, 1915.

MEMORANDUM FOR COMMANDING OFFICER, 27TH COMPANY.

Private Henry Clayton is indebted to the exchange $5 for credit extended.

RICHARD POE, Exchange Officer (2 copies).

Copy received,

JOHN DOE, Commanding 27th Company (1 copy).

FORM C.

MARINE BARRACKS, NAVY YARD.

New York, July 1, 1915.

31ST COMPANY, CREDIT ROLL, FOR JULY, 1915.

Rank.	NAME.	Pay.	Credit asked.	Credit approved.	Credit extended.	Remarks.
Pvt.	Brady, John	$15	$10	$10	$10	
"	Carney, Patrick	18	10	5	5	
	JAMES SMITH Commanding (4 copies)					
	Received copy, Henry White, Exchange Officer (1 copy)					
	Checkage requested, Henry White " " (2 copies)					
	Received copy, James Smith, Commanding (1 copy)					

FORM D.

MARINE BARRACKS, NAVY YARD.

Norfolk, Va., July 31, 1915.

COLLECTION ROLL FOR JULY, 1915.

Rank.	NAME.	Amount due.	Paid Aug. 4.	Paid Aug. 5.	Unpaid.	Remarks.
Pvt.	Brown, Arthur J.	$5.00		$5.00		
"	Fox, John	3.00	$3.00			
"	Gordon, William E.	5.00	4.00		$1.00	
	TOTAL	$13.00	$7.00	$5.00	$1.00	

JAMES WHITE, Exchange Officer.

FORM E.

POST EXCHANGE, MARINE BARRACKS.

$2.00 Norfolk, Va., Jan. 14, 1915.

I have received from the Exchange Coupon Book No. 107 and acknowledge myself justly indebted to the exchange for the value thereof, two dollars.

Private, U. S. M. C.

FORM F.

POST EXCHANGE, MARINE BARRACKS.

Norfolk, Va., Apr. 10, 1915.

SALES SLIP.

Quantities.	ARTICLES.	Amount.	
1	Pkg. Cigarettes		05
1	Box Blacking		05
2	Cakes Ivory		08
	Approved: John Wilson, Commanding. (Only when man is not in good standing.)		
	TOTAL		18

Received the above articles for which I acknowledge myself justly indebted to the post exchange.

Name............................

Rank....................

FORM G.

CASHBOOK-JOURNAL.

DEBITS. CREDITS.

Month of ...., 192...

Amusements.	Coupons.	Property.	Expense.	Loss and gain.	Shoemaker.	Tailor.	Barber.	Bills receivable.	Petty ledger.		Merchandise.	Cash.	Folio.	Date.	Council.	REMARKS.	Folio.	Invoice No.	Voucher No.	Check No.	Cash.	Merchandise.	Petty ledger.		Bills receivable.	Barber.	Tailor.	Shoemaker.	Loss and gain.	Interest and discount.	Property.	Coupons.	Amusement rooms.
									Accounts receivable.	Accounts payable.													Accounts payable.	Accounts receivable.									
																TOTALS.																	

FORM H.

Post Exchange, Marine Barracks, Norfolk, Va., Jan. 13, 1915.

STEWARD'S DAILY REPORT.

CASH.			MERCHANDISE.		
On hand last report. Since received.			On hand last report. Since received.		
Total. Turned in.			Total. Sales. Expended.		
Remaining.			Remaining.		

COUPONS.

On hand last report. Since received.			Outstanding last report. Since issued.		
Total. Issued.			Total. Redeemed.		
Remaining.			Outstanding.		

SALES AND SERVICES.

	Cash.		Coupons.		Credit.		Total.	
Merchandise: Store. Services, etc.: Barber. Tailor. Shoemaker. Laundry. Amusement rooms.								
Grand total.								

Cash-register reading:

-------------------------------,
Exchange Steward.

Examined and approved.

-------------------------------,
Exchange Officer.

FORM I.

Inventory, Exchange, Marine Barracks, Mare Island, Calif.

Date July 31, 1915.	Counted by Sgt. Wood	Chkd. by Lt. Smith.
Account Mdse.	Entered by Cpl. Jones	Chkd. by Lt. Udell.
Department..................	Priced by Lt. Meyer	Chkd. by Lt. Manney.
	Extended by do	Chkd. by do
	Totaled by Lt. Udell	Chkd. by Lt. Smith.

Quantity.	ARTICLE.	Unit cost price.	Cost value.	Unit selling price.	Selling value.	REMARKS.

....

....

....

Committee (last page)

N. M. C. 159—A&I.

FORM J.

BALANCE SHEET.

---, -------------------, 192

(Exchange.)

ACCOUNTS.	TRIAL BALANCE. Dr. footings.		TRIAL BALANCE. Cr. footings.		LOSSES.		GAINS.		ASSETS.		LIABILITIES.	
Amusements,	278	25			278	25						
Amusement rooms,			342	50			342	50				
Bills receivable,	61	15	49	25	*12	94			11	90		
Cash,	7,134	12	2,818	46			**	59	4,315	66		
Coupons,												
†Exchange,			11,666	22								
Expense,	127	13			127	13						
Interest and discount,			12	71			12	71				
Loss and gain,	*79	77	**	59								
Merchandise,	8,970	99	3,789	75	*34	47	539	00	5,720	24		
Property,	2,456	97	24	36	*24	36			2,432	61		
Barber,	31	60	129	25			97	65				
Shoemaker,	64	22	84	15			19	93				
Tailor,	58	90	64	98			6	08				
Accounts payable,	2,831	17	3,149	10							317	93
Accounts receivable,	257	40	220	35	*8	00			37	05		
	22,351	67	22,351	67								
†Exchange balance,			11,666	22								
Net gain or loss,			533	31	533	31						
Present worth,	12,199	53									12,199	53
	12,199	53	12,199	53	1,018	46	1,018	46	12,517	46	12,517	46

Total ration strength for period covered by this report-------------------------

Profits subject to distribution--

(If none, so state.)

Accounts payable will probably be settled-----------------------------------

To THE ADJUTANT AND INSPECTOR,

U. S. Marine Corps, Headquarters, Washington, D. C.

NOTE.—The amounts posted to the Loss and Gain account from the other accounts during the period covered will be entered in the columns "Losses" and "Gains" as separate items opposite the accounts concerned.

(OVER)

SCHEDULE OF BILLS RECEIVABLE.

NOTE.—Also comply with Art. 1217 (5) Marine Corps Manual

NAME.	RANK.	AMOUNT.	DATE DUE.
MEMORANDUM OF COST PRICE OF MERCHANDISE:			
Returned to dealer			
Donated			
Transferred to expense...........			
Transferred to property			

Profits distributed during the period, as follows:

.... $
.... $
.... $
.... $

ATTENDANTS AND EMPLOYEES.
(Including barber, shoemaker, etc.)

NAME.	POSITION.	Monthly compensation from exchange.

........................., 191

I CERTIFY that there is $........ on deposit in the................. National Bank of................. this date to the credit of the exchange; and cash on hand amounting to $.........................

....
Exchange Officer.

We certify that we have counted the cash on hand; examined the check book, bank book, and statement rendered by the bank for the period covered by this audit. We have carefully inventoried the merchandise and have audited the accounts of this exchange as required by art. 1215, Marine Corps Manual, and found them to be correct.

------------------------ ------------------------
------------------------ ------------------------
------------------------ ------------------------

Forwarded, approved.

Commanding.

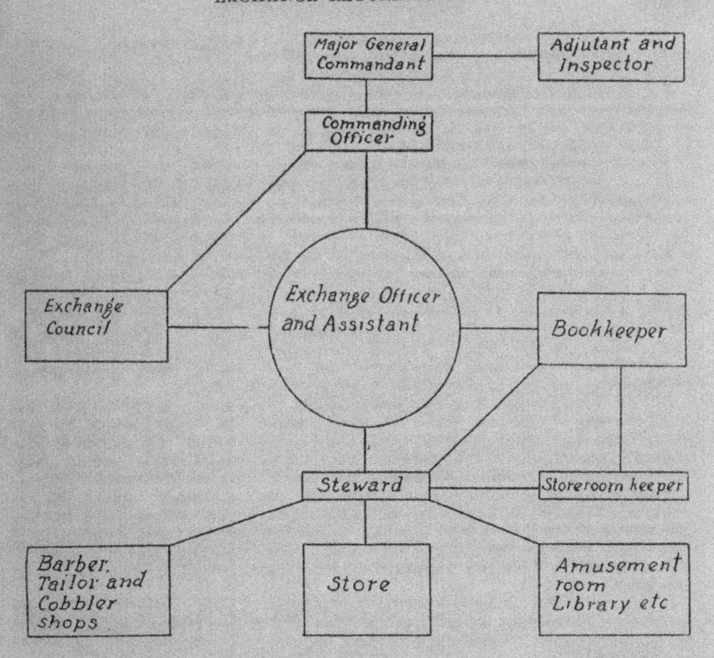

COMMANDING OFFICER (see Art. 1204).—Liaison with Headquarters, U. S. Marine Corps, exchange council, and exchange officer.

EXCHANGE COUNCIL (see Art. 1207).—Advisory body; liaison with commanding officer and exchange officer.

EXCHANGE OFFICER (see Art. 1205).—General manager; liaison with commanding officer, exchange council and all activities, either direct or through steward.

EXCHANGE STEWARD (see Art. 1206).—Accountable for all merchandise and property; direct supervision of all branches and activities; liaison with exchange officer and all departments and branches.

STOREROOM KEEPER (see Art. 1214 (16)).—Responsible to steward for all merchandise and property received in storeroom; keeps steward informed of any excess or deficiency; keeps prescribed record of all merchandise, etc., passing through his hands.

BOOKKEEPER (see Arts. 1206 (5) and 1214 (1-8)).—Liaison with exchange officer, steward, and storeroom keeper; keeps all accounts and records not required by regulations to be kept by other persons.

APPENDIX.

1222

(1) It being realized that there is need for a systematic method of entering accounts in the post exchange work, the following comments give in elementary form those things which will start the exchange bookkeeper out with the fundamentals of double-entry bookkeeping. Some of the outstanding faults heretofore apparent in these matters are:

1. Incompleteness—partial treatment of the more complex accounts.
2. Over-conservatism—adherence to old-fashioned theories.
3. Too much variety of method in various exchanges.
4. Over-development of system, leading to unnecessary clerical work.

(2) The term "double-entry" means simply that each transaction is recorded in full on each of the two sides of the records.

(3) The left-hand side is the *debit* side and the right-hand side the *credit* side.

(4) Since each transaction is entered in full on each side, it necessarily follows that after this is completed for one or any number of transactions, the total of all entries on one side equals the total of all entries on the other side, or, in other words, the books balance.

(5) The double-entry system of keeping books is subject to numerous variations, depending upon the character, scope, and object of the business. For exchanges, wherein the object and activities are limited, and the personnel as a rule comparatively lacking in technical knowledge of accounting, to say nothing of purely temporary employment therewith, it is considered advisable that a uniform and simple method be employed for keeping the records.

(6) In the double-entry system, the ledger is the book of *final entry*, and this book is an accurate, although condensed and tabulated, record of the business during its existence. It is from the ledger that the balance sheet is prepared.

(7) The cashbook-journal is the chronological record, in sufficient detail, of each transaction that occurs. It is in the cashbook-journal and the ledger that the double-entry principle is employed, the other records being in the nature of supporting records to the entries in the cashbook-journal.

(8) As above stated, each transaction must be recorded in full on each side of the books. This may, and often does, require the transaction to be split into two or more parts on either side, e. g., a cash receipt totaling $50.00 may emanate from the following sources: Merchandise sales, $30.00; barber shop, $8.00; tailor shop, $10.00; amusement room, $2.00. The entries in the cashbook-journal would show on the debit side in the column headed "*cash*," the one entry of $50.00; on the credit side in the various columns appropriately headed, the amount received from each source. In like manner, any other transaction is journalized to show the source of all money, merchandise, property, or services, or, on the other hand, the disposition of any money, merchandise, property, or services.

(9) In each case, when any of the above items is *received by* the exchange, an entry of the amount is made on the *debit* side of the cashbook-journal under the proper column. (This entry records the fact that something of value has been received.)

(10) At the same time there must be made in the *credit* side and in the proper column or columns an entry of equal amount. (This entry records the source of the item received.)

(11) In the same manner, when anything representing value is furnished or *paid out* by the exchange, an entry is made on the *credit* side of the cashbook-journal. (This records the fact that the exchange has actually given out something of value.)

(12) At the same time there is entered on the *debit* side of the cashbook-journal, in the appropriate column or columns, entries totaling the same amount. (This records the person, firm, or activity to whom the item was paid or given.)

(13) In making entries in the cashbook-journal all entries relating to a single transaction must be made on the same line. In this connection it must be borne in mind that *receiving* anything and *paying* for it are *two distinct transactions*, even though payment is spot cash.

Closing the Ledger.

1223

(1) *Classification of accounts in closing a ledger.*—In closing a ledger all accounts are divided into two classes: *Real accounts and fictitious accounts.*

(2) A *real account* is one which, as its name indicates, represents some tangible value, either as an asset or liability. Some of the *real accounts* kept

by an exchange are: *Cash, property, merchandise, bills receivable, accounts payable*, etc.

(3) A *fictitious account* is one which has no real value itself but serves to indicate a gain or loss on some account in the other classification, e. g., loss and gain, exchange, etc.

(4) As thus classified, *real accounts* are closed "*to*" or "*by*" balance as the case may require. *Fictitious accounts* are closed "*to*" or "*by*" loss and gain.

(5) An exception to the above rule is in the case of a merchandise account, which is a real account; but since a profit or loss is expected in this account, it is closed either "*to*" or "*by*" loss and gain, after the inventory for the month has been duly credited.

(6) The first step to be taken in closing a ledger is to open an account with "*loss and gain*," provided that this account has not been previously entered. The next step is to open an account with "*balance*," which will ultimately show on the debit side all resources or assets and on the credit side all liabilities. The *loss and gain* account will show, on completion, the losses on the debit side and the gains on the credit side. The differences between the debit and credit side of the loss and gain account will show the net gains or losses. During the month a loss and gain account furnishes a comfortable place for various items of loss and gain which can not be properly posted to other accounts. In closing the ledger it furnishes a place to put all the transfers from *fictitious* accounts and in this way show all the losses and gains in a single ledger title. The net gain or loss is then transferred to the exchange account and the "*balance*" of this exchange account will agree with the *present worth* shown in the *balance account*.

(7) *The use of ink in closing the ledger.*—Since the ledger's appearance largely depends on the proper use of inks, in its closure, it is highly important that the *red* and *black* ink used should always be employed properly and neatly.

(8) A general rule to follow in the use of ink is: A *red ink* entry is an entry that *never* indicates an actual transaction. Closing entries are made in *red ink* to distinguish them from the other entries in the *ledger*. It also indicates that the entries are to be transferred, either to another account or to the line below the rulings in the same account on the opposite side.

(9) *The rulings* of an account are made in *red ink*.

(10) Before closing a ledger, there are certain preliminary steps to be taken:

Take an inventory of all cash, merchandise, and property.

Take a trial balance.

(11) The *general ledger* accounts should appear in the same order as shown on the *monthly balance sheet*. The first account we will likely encounter, then, is the *amusement account*. Since this is an account involving losses or gains, we will close it in *red ink* "*to*" or "*by*" *loss and gain*. Assuming that the only transaction appearing on the ledger in the amusement account was a debit item—To merchandise, $100.00—we would know from this item that the exchange had donated to the men of the command certain articles of merchandise and the item is therefore a loss. We will then close the account "By *loss and gain*, $100.00." The account may then be ruled off.

(12) The $100.00 red-ink item is then transferred with black ink to the *debit* side of the *loss and gain* account and there shows up as a loss. All other accounts involving losses or gains are transferred in like manner. The accounts involving gains will, of course, be closed in red ink on the debit side of their respective accounts and transferred in black ink to the credit side of the *loss and gain*. Completing the *loss and gain* items, we now take up the *real accounts* (those closing to balance). These accounts are closed in red ink with the wording "*to*" or "*by*" *balance*, and the black-ink entries are then made to the opposite side of the *balance account*, showing-up in that account as an asset or liability. The account which we have just closed with the red-ink entry will then be ruled off and the balance appearing will then be brought down on the opposite side in black ink.

(13) The merchandise account is handled in the following manner: Assuming that the *debit* side of the *merchandise* account totals $2,552.00, this would mean that the prior inventory plus the receipts of merchandise in that month totaled the amount stated. The credit side shows that we disposed of $1,407.00 in merchandise. The inventory shows, we will say, $1,560.00. This is entered on the credit side in red ink. The total of the credit side is now $2,967.00; since the amount sold plus that now on hand is greater than that received and on hand the first of the month, it indicates that a gain has been made, and we will close the account with a debit entry of "*to loss and gain* $415.00." This red-ink entry is then carried to the credit side of the *loss and gain* account. The merchandise account is then ruled off, and the inventory balance is then brought to the debit side of the merchandise account below the rulings, in black ink.

(14) When all *loss and gain* items have been properly transferred and the balance entries properly made, we are now ready to close up the *exchange account*. We will assume that the debit side of the *loss and gain* account totals $294.55, showing that the total losses through the month total that amount. The credit side shows $490.27, which represents the total gains. A red-ink entry is then made on the debit side of enough to balance the account, which in this case would be $195.72. The entry is made "*to exchange*, $195.72." The amount is then carried to the credit side of the *exchange account* in black ink.

(15) Assuming that this credit side of the *exchange account* shows the present worth at the first of the month to be $2,768.81, the amount of $195.72 added to it will give the present worth now to be $2,964.53. The exchange account is then closed in red ink on the debit side "*to balance*, $2,964.53." The *balance account* is now the only account open on the books. We assume that the debit side shows an amount of $3,760.03, which represents the assets of the exchange. The credit side shows a total of $795.50, which represents the liabilities of the exchange. If the exchange *present worth* of $2,964.53 is now added to the credit side of the *balance account*, this account will now balance, and the ledger is properly closed.

(16) The *balance account* now shows us, in condensed form, an analysis of the *real accounts* and the *loss* and *gain* and analysis of the *fictitious* accounts.

(17) The next step is to prepare the *balance sheet*, which is required to be made up monthly. Since the data for this sheet are found in the *trial balance*, and in the *loss* and *gain*, and in the *balance account*, very little comment is necessary on the proper preparation, except to say that it should be the highest aim of the one preparing the sheet to be correct and neat in the drawing up of this sheet.

(18) *Taking a trial balance*.—After having posted all items into the ledger the next important step will be to show how the ledger is proved to be in balance; i. e., whether the amount of all the debits is equal to the amount of all the credits, or whether the amount of all its debit differences is equal to the amount of all its credit differences. The making of this test is called *taking a trial balance*, and the statement which results from making such a test is called the *trial balance*.

(19) The trial balance is so called because it is a test to see whether the ledger is in balance, after all the postings have been made therein, at any given time.

(20) Two forms of trial balances are in use—one called a *trial balance, by sums* and the other a *trial balance by differences*.

(21) A *trial balance by sums* (the one required in exchange balance sheet) is that form of test which shows whether the debit sums in the ledger are equal to the credit sums in the ledger. This form of the *trial balance* is based upon the principle of equality of debits and credits, which is characteristic of double-entry and which is supposed to constantly exist in the ledger.

(22) Since the debit and credit, or debits and credits, of each entry are exactly equal in amount to each other, it follows that the same equality will exist in the ledger if they are properly posted to it, and consequently that the

amounts of the debits and credits of all the ledger accounts, if properly entered in the trial balance, will also exactly agree in amount when the debit and credit columns of the trial balance are footed up.

(23) *Trial balance by differences*, not being used in exchange balance sheets, is not explained.

(24) *Work to be done preparatory to taking a trial balance.*—Commencing with the first account in the ledger, carefully foot up both sides of the account, care being taken to write the footings with a sharp-pointed lead pencil in small, light, plain figures close up to the last posting on each side of the account, so that such figures will not interfere with subsequent postings and may be easily erased, if desired, without defacing other figures. This done, so do, in turn, with all the other accounts in the ledger that require addition.

(25) The trial balance is required to be taken monthly, or at any other time desired. Aside from the fact of the requirement to execute a monthly balance sheet, another important object is attained, as in case the ledger does not balance at the end of any given month only one month's posting to the ledger will require to be reviewed in order to discover the error or errors made.

(26) *What the trial balance proves when it balances, and what it does not prove.*—When the trial balance balances, it simply proves that the ledger is in balance; i. e., that the amounts of all the debits and credits in the ledger accounts are equal. Although it is generally accepted as satisfactory evidence that all previous work has been correctly done when the trial balance balances, yet it is not *positive* proof that such is the case. This is due to the fact that various errors may be in prior work, which will not prevent the trial balance from balancing, such as a wrong or transposed entry; as merchandise (*debit*) and cash (*credit*), instead of cash (*debit*) and merchandise (*credit*). Or perhaps it may occur in case of an entire entry not being posted, or posting an item to the *right* side of the *wrong* account.

(27) The bookkeeper should bear these important facts in mind, to take the greatest possible care in doing *all* of his work, to the end that it may be done correctly.

(28) *Certain procedure when trial balance does not balance.*

(*a*) See that all additions were correctly made.

(*b*) Find the exact difference between the debit and credit footings of the trial balance, then see if there is not an amount in some entry in books or vouchers from which postings were made that exactly corresponds to the error, as such an amount may have been omitted in posting, or may have been posted twice to the same side of an account.

(*c*) If the error still exists and is exactly divisible by 2, see if there is not an amount in original record that corresponds exactly to the half difference, since such an amount, if posted to the wrong side of its own or some other account, would cause the disagreement.

(*d*) If the error is still undiscovered and the exact difference is exactly divisible by 9, 99, or 999, error may be a transplacement or transposition of figures, as 41, posted as 14, would cause a difference of 27, which is divisible by 9, produces a quotient of 3, which is the difference between 1 and 4, the figures transposed. All that will be necessary to check up an error of this sort is to find a posting whose difference in the tens and unit column is 3, and check back to book of original entry to find the entry which was transposed.

(*e*) Errors divisible by 9 may occur by a transposition in the hundreds or thousands columns also, or may occur in what is called a "*slide*"; i. e., $93.00 written as .93 causes an error of $92.07, which is, of course, divisible by 9.

CHAPTER 13.

SEPARATION FROM THE SERVICE.

DEATHS.

1301

Telegraphic reports.—Commanding officers of Marine Corps posts, recruiting stations, and other organizations will report immediately to the Secretary of the Navy by telegraph all deaths of officers or enlisted men under their command, giving name, rank, organization, date, place, cause of death, and information as to whether next of kin has been notified. Deaths of *accepted* applicants will be similarly reported.

M624

1302

Deaths under discreditable conditions.—Should a member of the Marine Corps die under discreditable circumstances, for instance, as the result of his own misconduct, it is directed that, in the telegram of notification to the family of the deceased, the mere fact only of his death in addition to the time and place, and the funeral arrangements, if any, be given, in order to spare the family any additional shock.

M402

1303

(1) *Flags to accompany bodies.*—Commanding officers of Marine Corps posts and stations are authorized to issue flags to accompany all bodies of officers or enlisted men whose deaths occur while in the service of the United States Marine Corps, forwarded or delivered to the next of kin or relatives for private interment, in order that the flags may be available for use at time of burial. Requests for such issue shall be construed as included in application for the body.

M577

(2) When funeral services are conducted by the Marine Corps, commanding officers are authorized to issue flags used for draping the coffins when requested. In cases of doubt as to whether the persons making requests are legally entitled to the flags, the requests should be forwarded to the Major General Commandant for authority prior to issuing the flags.

(3) Flags used for draping coffins at funerals where no next of kin is present will be preserved and stored for safe keeping, tagged with the name of the deceased and date of funeral, and held for a period of three months pending the receipt of request therefor from the next of kin. If at the end of three months no request is received the flag will be returned to store for reissue.

(4) The order of the commanding officer in writing will be sufficient voucher for dropping from the property account flags issued in accordance with the above.

DISCHARGES.

1304

(1) ***Classes.***—Discharges of enlisted men of the Marine Corps are divided into the following classes:

Class 1.—Honorable discharge.

(*a*) Upon expiration of enlistment or extended enlistment, or enrollment (fixed period or duration of war).

(*b*) Upon report of medical survey, disability in line of duty.

(*c*) Upon report of medical survey, disability not in line of duty, not due to own misconduct.

(*d*) On account of dependency of relatives, for convenience of the Government (death or disability of member of family occurring after enlistment and creating complete dependency).

(*e*) On account of dependency of relatives, for man's own convenience.

(*f*) To accept appointment in commissioned or warrant grade, or at Naval Academy, etc.

(*g*) For man's own convenience.

(*h*) For convenience of the Government.

(*i*) For inaptitude (including unfitness), involving no reflection upon moral character or conduct.

(*j*) By reason of under-age enlistment. (See subparagraph (5).)

Class 2.—Discharge.

(*k*) Upon report of medical survey, disability not in line of duty, and due to own misconduct.

(*l*) As undesirable; by reason of—

(1) Desertion.

(2) Habits and traits of character.

(3) Fraudulent enlistment.

(4) Conviction by a civil court.

(5) Other grounds.

(*m*) By reason of under-age enlistment. (See subparagraph (5).)

Class 3.—Bad conduct discharge.

(*n*) In pursuance of sentence of a summary or general court-martial.

Class 4.—Dishonorable discharge.

(*o*) In pursuance of sentence of general court-martial.

(2) ***Discharge gratuities.***—Men discharged under (*a*), except reservists, are eligible for honorable discharge gratuity if reenlisted within four months. Men discharged under (*a*), (*b*), (*c*), (*d*), (*h*), and (*i*) are entitled to the $60 bonus (if not previously received) and travel pay. Men discharged under (*e*), (*f*), (*g*), (*k*), (*l*), (*n*), and (*o*) are not entitled to $60 bonus or travel pay.

(3) ***Certificates used.***—In effecting the discharge of marines the following certificates will be used:

Class 1.—Honorable discharge.

Upon expiration of first enlistment, or extension thereof:

With award of good conduct medal, NMC 257a A&I (artificial parchment).

Without award of good conduct medal, NMC 257 A&I (parchment).

Upon expiration of reenlistment:

With award of good conduct medal, NMC 257a A&I (parchment).

Without award of good conduct medal NMC 257 A&I (parchment).

All honorable discharges from the Marine Corps Reserve, NMC 257g A&I (white paper).

Upon report of medical survey, disability in line of duty, NMC 257h A&I (parchment).

Upon report of medical survey, disability not in line of duty, not due to own misconduct, NMC 257h A&I (artificial parchment).

Class 1.—Honorable discharge—Continued.

All other discharges under Class 1, striking out in the fourth line of the certificate the words "the duration of the war," and substituting therefor the words "two years," "three years," or "four years," as may be proper, NMC 257i A&I (white paper).

Class 2.—Discharge.

In all discharges under Class 2, Form NMC 385a A&I (white paper).

Class 3.—Bad conduct discharge.

In all discharges under Class 3, Form NMC 385 (yellow paper).

Class 4.—Dishonorable discharge.

In all discharges under Class 4, Form NMC 385b (yellow paper).

(4) ***Preparation of certificates.***—Commanding Officers, in showing cause of discharge on discharge certificate, will in all cases employ the same phraseology as contained in examples under article 1309, but in no case will the discharge certificate of a marine discharge "upon report of medical survey for disability" bear notation indicating whether or not the disease or injury was incurred in the line of duty.

(5) ***Under-age discharges.***—Upon application of either of the parents or of the guardian of an enlisted man (other than an apprentice), who is under the age of 18 years at the time of receipt of such application, Marine Corps Headquarters will direct his discharge "by reason of under-age enlistment." He shall receive pay and the form of discharge certificate to which his service, after enlistment, shall entitle him, together with transportation in kind, in accordance with the Army appropriation act of June 30, 1921.

1305

Ceremony.—When a man is to be discharged upon expiration of enlistment with a good-conduct medal or bar the presentation of his discharge certificate and medal or bar will be made an occasion of ceremony whenever practicable.

1306

The character given to a man discharged for undesirability, inaptitude, or unfitness and entered on the face of the discharge certificate will be what his commanding officer thinks he deserves, and need not necessarily correspond with his markings for military efficiency, obedience, and sobriety.

1307

Medical survey.—Discharge of enlisted men upon report of medical survey for disability will not be effected until the man has been discharged from treatment in the hospital, except in cases of mental or incurable diseases.

1308

Before discharging a minor for any cause under circumstances where neither travel pay nor transportation in kind is to be furnished by the Government, the commanding officer will, if practicable, notify the parents or guardian of such minor a reasonable time in advance of the date set for discharge in order that they may send him funds with which to defray his expenses to his home.

1309

(1) ***Forwarding service-record books for discharge.***—The service-record books of men who are to be discharged upon expiration of enlistment within the United States will be forwarded to Headquarters, U. S. Marine Corps, or to Head-

quarters, Department of the Pacific, as the case may require; those for Headquarters, Marine Corps, to be forwarded not less than 15 days prior to expiration of enlistment from posts and recruiting stations east of the Mississippi River, and not less than 25 days prior to expiration of enlistment from posts and recruiting stations west of said river; service-record books to be forwarded to Headquarters, Department of the Pacific, not less than 15 days prior to expiration of enlistment.

(2) *Completing service-record books of discharged men.*—The service-record books of all men who are to be discharged for other causes will be retained at the posts (or on board the ships) where said men are to be discharged until the discharge certificates have been delivered, when an entry to the effect that the man has been discharged will be made in the place provided in the service-record book in accordance with one of the following examples, as may be appropriate:

(*a*) "Dishonorably discharged at (place of discharge) on (date of discharge), pursuant to the sentence of a general court-martial. Character: Bad.

"_______________________________, U. S. M. C.,
"*Commanding.*"

(*b*) "Discharged at (place of discharge) on (date of discharge), with a bad-conduct discharge, pursuant to the sentence of a summary (or general) court-martial. Character: Bad.

"_______________________________, U. S. M. C.,
"*Commanding.*"

(*c*) "Discharged at (place of discharge) on (date of discharge), as unfit for service (or undesirable), by order of the Major General Commandant (or, by direction of the Secretary of the Navy). Character: ______________

"_______________________________, U. S. M. C.,
"*Commanding.*"

(*d*) "Discharged at (place of discharge) on (date of discharge), upon report of medical survey, for disability. Disease or injury was (or was not) the result of his own misconduct. Character: ______________

"_______________________________, U. S. M. C.,
"*Commanding.*"

(*e*) "Discharged at (place of discharge) on (date of discharge), by special order of the Major General Commandant, for his own convenience (or for convenience of the Government). Character: ______________

"_______________________________, U. S. M. C.,
"*Commanding.*"

(*f*) "Discharged at (place of discharge) on (date of discharge), by reason of underage enlistment. Character: ______________

"_______________________________, U. S. M. C.,
"*Commanding.*"

1310

Bad-conduct discharges.—When an enlisted man is sentenced by a summary court-martial to bad-conduct discharge, his accounts will be closed and transmitted for discharge at such time as date of discharge is decided upon, unless he is to be transferred to another station or ship for discharge, when his staff returns will be transmitted from his new station.

1311

Medical survey discharges.—When an enlisted man has been surveyed and recommended for discharge by a board of medical survey, his accounts shall be closed and forwarded to the pay officer carrying his accounts on the same date that the report of survey is transmitted, or as soon thereafter as practicable, but the man will not be discharged until an order for his discharge has been received from the Major General Commandant, excepting in cases of recruits under training at training stations and training camps, where the report of medical survey has been approved by the local authorities, as provided in article D-8030, Bureau of Navigation Manual, 1921.

1312

Address cards.—Commanding officers will furnish men about to be discharged with three or more cards (N. M. C. 684), to be used by them in reporting to Headquarters any change in their address for a period of three months after discharge.

1313

(1) *Medical survey discharge of recruits.*—In the cases of recruits under training, commandants and commanding officers of training stations and training camps are authorized to approve the recommendations of boards of medical survey recommending the discharge of enlisted men of the Navy in advance of the bureau's approval. When such action is taken in any case the medical survey showing the action taken shall be forwarded in the usual manner. In medical surveys on any men other than recruits the papers shall be forwarded to the bureau before final action is taken. (Art. D-8030, Bureau of Navigation Manual.)

(2) Commanding officers and pay officers concerned are hereby authorized to proceed immediately, upon approval of report of a medical survey by the local authorities, in case of a recruit under training, as provided in article D-8030, above quoted, with preparation of discharge certificate and final settlement, without awaiting orders, or copy thereof, respectively, from the Major General Commandant.

1314

Discharge of probationers.—In no case should a summary or general court-martial probationer be discharged upon report of medical survey until the bad-conduct or dishonorable discharge has been remitted by the Department.

1315

No transfers after accounts closed.—In order to avoid confusion and delay in final settlement, no transfers will be made or authorized after a man's accounts have been closed preliminary to discharge.

1316

In the case of a reservist, the Paymaster, Headquarters, Marine Corps, Washington, D. C., should be notified of the discharge by letter in triplicate.

1317

Orders for discharge of men who are surveyed on board ship, or while serving with expeditionary forces, or who join a marine barracks direct for discharge, will be issued from Marine Corps Headquarters as soon as practicable after receipt from the Bureau of Medicine and Surgery of approved report of medical survey. That there may be no delay in these cases, commanding officers will promptly report their joining.

1318

(1) *Preparation of discharge certificates.*—Except in the case of a man to be discharged in the Department of the Pacific or any other place where specific authority to prepare discharge certificates and to award good-conduct insignia has been or may be given by the Major General Commandant, all parchment and imitation parchment discharges will be prepared at Marine Corps Headquarters, and the certificate of good-conduct award thereon, if any, signed by the Major General Commandant, the certificate being forwarded for completion and delivery to the place where the man is serving.

(2) In the case of a white or yellow discharge certificate, such certificate will be prepared, as well as delivered, at the place where the man is serving, the discharge, and the delivery certificate thereon, being signed by the commanding officer of the post.

(3) Final settlements on discharge will be signed in a similar manner. Delivery of check will invariably be made by a commissioned officer.

1319

Transfer for discharge.—Where marine detachments on shore are in charge of noncommissioned officers, and in cases of marine detachments serving on board ships in home waters, the marine will be transferred without specific orders from the Major General Commandant to the nearest Marine Corps post commanded by a commissioned officer, for discharge, at least two weeks prior to the expiration of enlistment.

1320

Final settlements.—The Major General Commandant (or the commanding general Department of the Pacific) will send to the pay officer having the man's accounts a copy of the order sent to the commanding officer to prepare a discharge certificate. Upon receipt of the copy of the order for discharge, the pay officer will make final settlement, including therein, in addition to the credit or debit for undrawn or overdrawn clothing theretofore certified by the post quartermaster or office keeping the clothing account rolls, the additional amount according to the prescribed scale of allowances for clothing due to the date of discharge, such date to be determined by the paymaster according to the time necessary to allow for the receipt of the final statement and check (if any) at the place of discharge. The officer delivering the discharge certificate will in all cases enter therein the amount paid.

1321

Pay and clothing accounts of men to be discharged, upon expiration of enlistment, made out to include the day of discharge, will be forwarded to the proper paymaster at such time prior to the expiration of such enlistment as will enable the paymaster to make out and return final statements. The paymaster will make out the final statement, and will forward such statement with check, without further notification, to the man's commanding officer in time to reach him prior to the expiration of his enlistment.

Discharge While on Foreign Shore Duty.

1322

(1) *Waivers to consular aid and transportation.*—Marines who are to be discharged while on shore duty on foreign stations shall not, in consideration of being retained at such stations for discharge, be required to execute waivers

of claims to consular aid and to transportation and subsistence for sea travel incident to return from place of discharge to place of current enlistment.

(2) Waivers of claims to consular aid and transportation will only be required in cases of marines who are to be discharged upon their own request on expiration of enlistment from a ship on foreign station, in accordance with article 1689 (2), Navy Regulations, 1920.

(3) Waivers in such cases will be worded as follows:

"I, ——— ———, serving as a marine on board the U. S. S. ———, request to be discharged upon expiration of my enlistment, ———, 19—, in the port of ———. If so discharged, I hereby waive all claims to consular aid and to transportation to the United States."

(4) Waivers will be executed in duplicate and one copy forwarded to Headquarters, United States Marine Corps, through official channels; the other will be securely pasted in the man's service-record book.

(5) *Transportation and subsistence.*—Upon the discharge while on foreign shore duty of any enlisted man with an honorable discharge, the law requires that transportation and subsistence in kind for the sea travel over the statutory shortest usually traveled route to his place of enlistment ~~or bona fide home or residence, whichever is elected~~, should be tendered him. If this tender be accepted, then this transportaton and subsistence in kind should be supplied the discharged man after his discharge. For the land travel in the journey, both within and without the United States, mileage at the rate prescribed by law rather than transportaion and subsistence in kind should be credited him in his final settlement. If the man declines to accept the tender of the transportation and subsistence in kind for the sea travel, through his desire immediately to reenlist, or for any other cause, a proper record of this tender and of its nonacceptance, or its waiver, should be made as required by instructions contained in the Paymaster's Manual, Marine Corps, 1921, but in a case where the accounts are paid by a naval pay officer this record should further be made under "Remarks" on pay rolls where the final statement of the account is extended.

c)

1323

Undesirable discharges.—When requests are made for the discharge of a man from the service as "undesirable" a full report of the circumstances will be made, together with the recommendation of the company or detachment commander, the recommendation of the post commander, and a statement of the man concerned, or a declaration that he desires to make no statement. Requests for discharges of this nature will not be made by telegraph. No discharges of this nature will be effected without specific authority from the Major General Commandant, or, in the Department of the Pacific, the departmental commander.

1324

Although the Navy Regulations authorize the discharge of marines because of undesirability, recourse to such action should be had only in exceptional circumstances, i. e., where the man concerned is constantly committing breaches of discipline of such minor character as not to warrant trial by court-martial, or who, because of ~~lack of intelligence~~, uncleanliness, immorality, etc., is clearly undesirable for retention in the service. Discharge as undesirable is not to be used as a punishment, but rather as a means of disposing of persons who are useless to the service.

1325

Closing accounts for settlement.—Immediately a man is recommended for discharge, a statement of his account for settlement on discharge (N. M. C. 90) will be submitted by the officer keeping the pay rolls to the proper paymaster of the Corps, or of the Navy, in which statement there will be credited or debited the amount then due or overdrawn on account of clothing. A settlement of the clothing account on N. M. C. 146, certified by the post quartermaster or other officer keeping the clothing account roll, supporting this sum, must accompany N. M. C. 90. Where accounts of marines on shore or sea duty are carried by naval paymasters, N. M. C. 146, prepared as above indicated, but extended to include the date of discharge, will be furnished such paymasters.

1326

(1) When closing accounts for settlement on form N. M. C. 90, for discharge "upon report of medical survey, for disability," the form will contain, in brackets, immediately after the cause for discharge, a notation as to whether or not the disease or injury was due to the man's own misconduct.

(2) The entry in the man's service-record book will also show whether or not the disease or injury was the result of his own misconduct.

Removal of Insignia Upon Discharge for Bad Conduct, Etc.

1327

(1) The attention of commanding officers is directed to that portion of Navy Department General Order No. 110, revised July 10, 1916, which requires the removal of all distinctive insignia of uniform prior to the discharge of enlisted men in all cases of bad conduct, dishonorable, and undesirable discharges.

(2) Inasmuch as the requirements of General Order No. 110, revised July 10, 1916, are applicable to the Marine Corps it is directed that the following procedure govern in cases of enlisted men of the Marine Corps.

(3) Prior to the discharge of a marine for bad conduct or undesirability (pursuant to the sentence of a general or summary court-martial by order of the Secretary of the Navy or the Major General Commandant, etc.) all Marine Corps bronze or brass buttons, cap ornaments, figures and letters (hat), chevrons, trousers' stripes, service stripes, rating badges (gun pointers, gun captains, musicians) will be removed from all articles of uniform clothing in his possession at date of discharge.

(4) Buttons so removed will be replaced by suitable plain commercial buttons, which will be supplied by the quartermaster's department upon requisition on the proper depot quartermaster. It is proposed to supply buttons permitting of quick attachment without sewing, i. e., similar to the commercially known "bachelor's button." These buttons will be supplied gratuitously to the man about to be discharged and will be dropped from the returns of the accountable officer in such manner as may be prescribed by the Quartermaster.

(5) Cap ornaments, figures, and letters (hat), removed as above directed, will be taken up by the post quartermaster. When buttons have accumulated in sufficient quantities they will be invoiced to the nearest depot quartermaster.

(6) Assuming that chevrons, service stripes, etc., would ordinarily be removed in a more or less worn and soiled condition and unsuitable for reissue, such insignia may be destroyed immediately.

(7) Medals of honor and medal of honor rosettes, distinguished service medals, navy crosses, good-conduct medals and bars, Dewey medals, West Indian campaign medals and Special meritorious service medals, life-saving medals, campaign badges (Civil War, Spanish, Philippine, China, Army of Cuban Pacification, Nicaraguan, etc.) and ribbon bars in connection therewith, brevet medals, and target insignia (distinguished marksman's badges, Marine Corps competition medals, Marine Corps division competition medals, Marine Corps division competition pistol medals, expert riflemen's badges, sharpshooter's badges, and

marksman's pins, issued by the Marine Corps, distinguished pistol-shot's bars, expert rifleman's bars, and sharpshooter's badges, issued by the Navy, etc.), are the private property of the men to whom awarded and will be retained by them, excepting good-conduct medals forfeited by sentence of a general court-martial. A man discharged for bad conduct or for undesirability should not be permitted, however, to wear any of these insignia upon his departure from the post or ship.

(8) As marines dishonorably discharged pursuant to sentences of general courts-martial surrender their uniform clothing in exchange for civilian clothing, the foregoing directions are only applicable in such cases so far as to require that precautionary measures be taken to see that men so discharged surrender all articles mentioned in paragraph (3) and that upon their departure they do not wear any of the insignia mentioned in paragraph (7) or any distinctive insignia of the uniform.

1328

RETENTION IN SERVICE TO WORK OFF INDEBTEDNESS.

(1) Marines who are to be discharged with a bad-conduct discharge, upon report of medical survey, or for other cause for the convenience of the Government, will not be held in the service for the purpose of working off indebtedness to the Government, except in cases in which such indebtedness has been incurred for the man's own benefit, such, for example, as for transportation from furlough to post.

(2) Marines who are to be discharged for their own convenience will not be discharged while in debt to the Government.

(3) In the event that any case arises where the above rules should not be applied, it will be reported to the Major General Commandant for special consideration.

RETIREMENT.

1329

(1) *Under Army laws.*—Enlisted men of the Marine Corps are entitled to retirement under the laws and regulations provided for the Army. The application of an enlisted man of the Marine Corps for retirement shall be made to the President of the United States, and in computing the thirty years necessary to entitle him to be retired all service in the Army, Navy, and Marine Corps shall be credited in accordance with Army Regulations.

(2) *Closing accounts for retirement.*—After approval of an enlisted man's application for retirement, an order shall be issued from Marine Corps Headquarters transferring him to the retired list. Upon receipt of such order by his immediate commanding officer a final statement shall be prepared, his pay and clothing accounts being closed (on forms N. M. C. 90 P. M. and N. M. C. 146 QM, respectively) to and including date of retirement and forwarded to the pay officer carrying his accounts. No discharge shall be given, however, but his name shall be dropped from the active list and his staff returns closed and forwarded to Headquarters. His post office address on retirement will also be forwarded.

(3) A descriptive list, Form N. M. C. 559 A & I, shall be made out by the officer transferring an enlisted man to the retired list and forwarded with a copy of the Major General Commandant's order directing the transfer, to the pay officer who makes final settlement with the marine, who will in turn forward it with the necessary transfer pay accounts to the Paymaster, Headquarters, Marine Corps.

(4) *Ceremony.*—When an enlisted man of the Marine Corps is to be placed on the retired list after 30 years' service, the presentation of his retirement papers will be made an occasion of ceremony, the scope of the ceremony and the number of troops participating being left to the discretion of the commanding officer of the post or station.

(5) *Pay.*—The authorized pay and allowances of retired enlisted men of the Marine Corps shall be paid them monthly by the Paymaster, Headquarters, Marine Corps.

(6) *Monthly report of address.*—On the last day of each calendar month retired enlisted men shall report to Marine Corps Headquarters their post-office address, and shall promptly report any change therein. Blank cards for this purpose shall be furnished retired enlisted men by Headquarters.

CHAPTER 14

MARINE CORPS RESERVE.

ESTABLISHMENT.

1401

(1) The Marine Corps Reserve is established and maintained under the provisions of the act of August 29, 1916, and subsequent modifications thereof.

(2) The Marine Corps Reserve is composed of citizens of the United States, who, by enrolling therein or by transfer thereto from the Marine Corps, obligate themselves to serve in the Marine Corps in time of war or during the existence of a national emergency declared by the President.

(3) Reference to officers in this chapter includes commissioned and warrant officers and pay clerks unless otherwise indicated.

CLASSES.

1402

(1) *Class 1*, Fleet Marine Corps Reserve, is composed of former officers and enlisted men of the Marine Corps and is subdivided as follows:

(*a*) Former permanent officers of the Marine Corps who have left the service under honorable conditions.

(*b*) Former enlisted men of the Marine Corps who have been honorably discharged from the Marine Corps after at least one four-year term of enlistment or an enlistment during minority. In case of more than one such enlistment, the discharge from the last enlistment must have been honorable.

(*c*) Enlisted men of the Marine Corps transferred to the reserve in lieu of discharge, on application approved by the Major General Commandant, at the expiration of a term of enlistment, who may be then entitled to an honorable discharge after 16 years' naval service, excluding inactive service in the Reserve.

(*d*) Enlisted men of the Marine Corps transferred, on application approved by the Major General Commandant, at any time after completion of 20 or more years' naval service, excluding inactive service in the Reserve.

(2) *Class 2* is composed of officers who have had two years' military experience as such and are not less than 20 or more than 35 years of age upon enrollment, and of men in enlisted ranks who have had military experience or have enrolled from the naval militia of any State, Territory, or the District of Columbia and who are not less than 18 or more than 35 years of age upon enrollment.

(3) *Class 3*. There is no Class 3 in the Marine Corps Reserve corresponding to the Naval Auxiliary Reserve.

(4) *Class 4* is composed of officers and men who are capable of rendering useful service to the Marine Corps or in connection with the Marine Corps in the defense and maintenance of naval utilities.

(5) *Class 5*, Marine Corps Reserve Flying Corps, is composed of:

(*a*) Officers, provisional, who qualify professionally by such examination as may be prescribed by the Major General Commandant.

(*b*) Officers, confirmed, who have enrolled after leaving the naval service under honorable conditions and who are qualified for aviation duties, who are surplus graduates of the Navy Aeronautical School and are appointed second lieutenants in the Marine Corps Reserve Flying Corps, and officers, provisional, who are confirmed upon qualification in their rank.

(*c*) Men, provisional, who qualify professionally by such examination as may be prescribed by the Major General Commandant.

(*d*) Men, confirmed, qualified for aviation, who have enrolled after being honorably discharged from the naval service after one or more four-year enlistments or after an enlistment during minority, and those who are confirmed upon qualification in their rank.

(*e*) Enlisted men of the Marine Corps who are qualified for aviation and are transferred from the Marine Corps to the Marine Corps Reserve Flying Corps, upon completion of 16 years' service, under the same conditions as Class 1 (*c*).

(*f*) Enlisted men of the Marine Corps who are qualified for aviation and who are transferred to the Marine Corps Reserve Flying Corps upon completion of 20 or more years' service, under the same conditions as Class 1 (*d*).

(6) *Class 6*, Volunteer Marine Corps Reserve, is composed of those officers and men who obligate themselves to serve in the Marine Corps in time of war or declared emergency, who do not receive retainer pay or uniform gratuity in time of peace, and are governed by the same rules as other classes in time of war.

ENROLLMENT.

1403

(1) *All members enrolled.*—All members of the Marine Corps Reserve are enrolled, except those enlisted men transferred to the Fleet Marine Corps Reserve and the Marine Corps Reserve Flying Corps upon the completion of 16 years' service or upon or after the completion of 20 years.

(2) *For four years.*—All enrollments in all classes are for four years and are for general service. All reservists are required to take the oath of allegiance to the United States.

(3) *Qualifications.*—Before enrollment candidates must furnish evidence satisfactory to the enrolling officer as to qualifications and character.

(4) *The physical requirements* for enrollment are the same as those prescribed for enlistment of men in the Marine Corps, with the modifications set forth in "Circular Relating to the Physical Examination of Applicants for Enrollment in the United States Naval Reserve Force."

(5) *Age.*—Former enlisted men who are over 41 years of age will not be enrolled in the Fleet Marine Corps Reserve without the prior approval of the Major General Commandant, and no person will be enrolled in any class of the Reserve who is 60 or more years of age.

(6) *Enrollment as officers* in any class of the reserve of applicants, who have had no prior service as officers in the Marine Corps or Marine Corps Reserve will not be effected without the authorization of the Major General Commandant.

(7) Enrollment and reenrollment of officers in the Fleet Marine Corps Reserve and in the Marine Corps Reserve Flying Corps will be effected only by authorization of the Major General Commandant upon application. Reenrollments of officers in confirmed rank in all classes of the reserve will be subject to the same authorization.

(8) *Reenrollments.*—The presentation of the letter of discharge or of the letter accepting resignation will be sufficient authority for the reenrollment of officers in Classes 2 and 4, subject to the general instruction issued from time to time.

(9) Applications for reenrollment as officers of former reservists who have had subsequent enlisted service in the Army, Navy, or the Marine Corps will be referred to the Major General Commandant for action in the light of records during the enlistment.

(10) Enrollments and reenrollments in Class 4 are subject to the approval of the Major General Commandant.

(11) *Naval Militia.*—A member of the Naval Militia of a State, Territory, or of the District of Columbia may enroll in the Marine Corps Reserve.

(12) *Disenrollment.*—With the exception of enlisted men transferred to the Fleet Marine Corps Reserve or the Marine Corps Flying Corps, and subsequently promoted to officers, the Major General Commandant may disenroll any officer who is not qualified to perform the duty for which enrolled. Commissions of members transferred to the Fleet Marine Corps Reserve or the Marine Corps Reserve Flying Corps, and subsequently promoted, may be revoked under similar conditions.

RANK, PROMOTION, AND REDUCTION.

1404

(1) *Ranks and grades authorized.*—The various ranks and grades of the Marine Corps up to and including that of major are authorized for the Marine Corps Reserve.

(2) *Rank upon transfer.*—Transfers of enlisted men of the Marine Corps to the Fleet Marine Corps Reserve, or the Marine Corps Reserve Flying Corps, upon the completion of 16 years' service or after the completion of 20 years' service will be with the rank actually held.

(3) *Provisional and confirmed rank.*—Enrollments in the Fleet Marine Corps Reserve are with confirmed rank. Other reservists upon first enrollment are assigned a provisional rank corresponding to the military knowledge and experience or to the skill in the technical ability which is expected to be of service to the Marine Corps. Those without military experience or technical ability will be enrolled as private, provisional.

(4) *Promotions to provisional rank or grade.*—All promotions (except of members of the Fleet Marine Corps Reserve, and of the Marine Corps Reserve Flying Corps, transferred from the Marine Corps after 16 or 20 years' service, who are promoted only while on active duty) are to provisional grade or rank, which is subject to confirmation. Promotions to a higher rank or grade of members of the Fleet Marine Corps Reserve and the Marine Corps Reserve Flying Corps on active duty who are transferred to these classes after 16 or 20 or more years' naval service are neither provisional nor confirmed, but carry the higher retainer and other pay of the corresponding rank in the Marine Corps.

(5) *Former permanent officers* of the Marine Corps may be appointed in the Reserve with confirmed rank, in the grade or rank held by them upon separation from the corps, without examination, other than physical. All other confirmations as officers in the Reserve will be subject to active service for three months in the provisional rank or grade, to physical examination, and to examination and recommendation by a board of three officers of the Marine Corps, not below the rank of major.

(6) *Confirmations* in grades or ranks other than commissioned will be subject to three months' active service in the provisional grade or rank, to examination and to certificate of commanding officer of qualification for the performance of duties.

(7) *Reductions.*—Members who enroll in or are transferred to the Fleet Marine Corps Reserve with noncommissioned rank can not be reduced from that rank except upon their own request or by sentence of court-martial.

(8) *Reenrollments are made in the confirmed rank,* if any, held upon discharge, otherwise with the provisional rank. In case of a reservist holding confirmed rank, and a higher provisional rank, he may be reappointed by Headquarters,

Marine Corps, to the higher provisional rank upon reenrollment in the confirmed rank.

(9) *Enrollments of duration of war* men will be made provisionally in the ranks held by them on discharge.

(10) *Enrollment and reenrollments in class 4* are subject to the approval of the Major General Commandant.

(11) *Promotion to warrant or commissioned grade.*—Members of the Fleet Marine Corps Reserve on active duty, who establish their qualifications to the satisfaction of the Secretary of the Navy, may be appointed pay clerks, quartermaster clerks, marine gunners, or second lieutenants in the Reserve.

(12) Reservists in an inactive status who have been transferred from the Marine Corps upon completion of 16 or 20 years' service do not forfeit ther retainer pay, but their active duty pay and clothing accounts in enlisted ranks are closed.

(13) *Signature to officers appointments.*—Appointments of officers in Classes 2, 4, 5, and 6, will be signed by the enrolling officer, by authority of the Major General Commandant. Appointments of officers in Class 1 will be signed by the Secretary of the Navy.

ADMINISTRATION.

1405

(1) *By Major General Commandant.*—The Marine Corps Reserve is organized and administered by the Major General Commandant, who holds the same relation to it as to the Marine Corps.

(2) *Commanding Officer inactive Reservists.*—The officers in charge of the Marine Corps Reserve, Headquarters U. S. Marine Corps, will be the Commanding officer of all inactive reservists.

(3) *Orders for active service for training.*—When authorized by the Major General Commandant, the commanding officer of the Marine Corps Reserve will, upon request for active service for training as a reservist, issue the necessary orders to the designated training center, furnishing the commanding officer of the center with a copy of the orders. He will also, in case the reservist is in an enlisted rank or grade, issue the necessary tranportation and subsistence. Except by authority of the Major General Commandant, a reservist will not be assigned to active duty for training for a longer period than two months during an enrollment.

(4) *Upon completion of the period of training,* the comanding officer of the training center will order the reservist to his home, furnishing the Adjutant and Inspector, Headquarters, Marine Corps, with a copy of the order. He will also, in case the reservist is in an enlisted rank or grade, issue the necessary transportation and subsistence.

(5) *Upon first reporting for active service,* for training or duty, reserve officers and men will be examined by a medical officer, who will make appropriate entry in the health record showing physical condition. Similar entry will be made upon transfer to inactive status.

(6) *Transfers between classes.*—Reservists may, upon approval of the Major General Commandant, transfer from one class to another in which qualified without prejudice to rank or confirmation, but in the case of a confirmed rank, the general nature of the duties must not be changed.

(7) *Assignment to inactive status.*—Reservists will be assigned to an inactive status upon enrollment, unless designated for immediate active service. Two months' active service for training are prescribed for reservists during enrollment, which may be performed in one or more periods as the reservist may elect.

(8) *Enlistment of reservists in other forces prohibited.*—A member of the Marine Corps Reserve is not permitted to enlist or enroll in the land or naval forces of any State, Territory, or of the District of Columbia.

(9) ***Reservists are required to report for inspection*** under such conditions as may be prescribed by the Major General Commandant.

(10) ***Applications for transfer to fleet.***—Commanding Officers are required to exercise particular care to forward applications of enlisted men for transfer to the Fleet Marine Corps Reserve or the Marine Corps Reserve Flying Corps, on the completion of 16 years' service, in time to permit the transfer in lieu of discharge.

(11) ***Endorsement of applications.***—The endorsement of commanding officers on the application of enlisted men for transfer to the Fleet Marine Corps Reserve, Class 1 (*c*) and 1 (*d*) upon completion of 16 or 20 or more years' naval service will contain the following information:

Class 1 (*c*)—16 years in lieu of discharge.

(*a*) Whether or not it is proposed to recommend good conduct medal insignia. (If recommended and it is necessary to alter the recommendation, Headquarters of the Marine Corps will be notified immediately.)

(*b*) The amount of time lost under N. R. 554 during current enlistment and the cause thereof.

(*c*) Any extraordinary heroism that occurred during current enlistment.

Class 1 (*d*)—20 or more years.

(*d*) The average mark in conduct during current enlistment.

(*e*) Any extraordinary heroism that occurred during current enlistment.

(*f*) If the applicant on the date he is due for discharge by expiration of his current has completed exactly 20 years naval service and is to be transferred to the Fleet Marine Corps Reserve, in lieu of discharge, there will be shown the amount of time lost under N. R. 554 during current enlistment, the cause thereof, and whether or not the applicant is recommended for good conduct medal insignia.

(12) ***Discipline.***—Members of the Marine Corps Reserve are subject to the same discipline as other members of the Marine Corps only when on active duty, and when traveling to and from their homes for active duty, except former enlisted men of the Marine Corps transferred to the Fleet Marine Corps Reserve or the Marine Corps Reserve Flying Corps, upon 16 or 20 years' service, who are at all times subject to that discipline.

(13) ***Leaving the United States.***—Reservists intending to leave the United States, or to change their residence or status in any way, which may render them unavailable for call, are required to report the fact and the probable duration of the absence from the United States, or the unavailability, to the Adjutant and Inspector. Disenrollment of enrolled members may be effected if the absence or other unavailability be sufficiently prolonged.

UNIFORM AND INSIGNIA.

1406

(1) ***Uniform issued upon reporting for active duty.***—Reservists, except officers, will have issued to them upon reporting for active duty for training or other purpose, such articles of uniform and clothing as may be necessary for the proper performance of duty, the value of which will be charged against the clothing gratuity and clothing allowance, and any excess against the pay. Commanding officers are charged with the responsibility of preventing loss to the Government by the reason of overissue.

(2) ***Good-conduct medals.***—Members of the Marine Corps Reserve who during the period between April 6, 1917, and November 11, 1918, performed active service in enlisted grades will for such service be awarded good-conduct medals under restrictions imposed by regulations and orders applicable to enlisted men of the Marine Corps.

(3) ***Marine Corps Reserve buttons*** will be issued to transferred and enrolled members of the Marine Corps Reserve.

(4) *Honorable discharge buttons* will be issued to discharged members of the Marine Corps Reserve.

DISCHARGE AND RETIREMENT.

1407

(1) *A reservist is entitled to discharge,* except in time of war or declared emergency, upon application to the Major General Commandant, or upon expiration of enrollment. If a reservist holding enlisted rank is discharged or disenrolled without compulsion before the expiration of enrollment, he is required to reimburse the Government for any clothing gratuity credited to him during enrollment.

(2) *Preparation of discharge certificate.*—The commanding officer of the Marine Corps Reserve will prepare certificates for the discharge of inactive reservists, upon expiration of enrollment or otherwise. The Paymaster will deliver these certificates together with the final settlement, and forward same direct to the address of the reservists.

(3) *Disenrollment for age.*—Enrolled reservists will be disenrolled upon reaching the age of 64 years. If in active service at the time of war or declared emergency upon attaining that age, reservists may be continued in an active status by the Major General Commandant, but in no case beyond six months after the war or national emergency shall have ceased.

(4) *Retirement.*—Enlisted men transferred to the Reserve after 16 or 20 years' service may be retired upon written application upon the completion of 30 years' service, counting inactive service after transfer. Other reservists are not entitled to retirement.

RECORDS AND REPORTS.

1408

(1) *Finger prints* of applicant for enrollment in enlisted grades of the Reserve will be taken and forwarded to the Adjutant and Inspector under the rules governing the recruiting service.

(2) *Change of address.*—Reservists are required to report changes of address to the Adjutant and Inspector. When practicable, prescribed blank forms will be used.

(3) *All enrollment papers* will be forwarded to the Adjutant and Inspector.

(4) *Report of injury.*—Reservists are required to report to the Adjutant and Inspector any serious injury or impairment of health, and any other conditions which may make them unavailable for a call to active duty

(5) *Service-record books.*—Commanding officers of the training centers will enter in the reservists' service-record books dates of joining and transfer, and markings for proficiency and conduct. The date of joining and transfer will be reported by letter in each case to the Major General Commandant, and three copies of each such letter will be forwarded to the Paymaster, Headquarters, Marine Corps.

(6) *Fitness reports* on reserve officers in an active duty status will be submitted under the same instructions as those applicable to officers of the Marine Corps on active duty.

(7) *The retainer pay accounts* of officers and men of the Marine Corps Reserve are carried by the Paymaster, Headquarters, Marine Corps. Reservists transferred from the Marine Corps after 16 or 20 years' service are paid monthly; all other reservists are paid annually. Checks are transmitted in accordance with rules established by the Paymaster.

(8) *Muster, pay, and enrollment cards.*—The enrolling officer, in case of enrollment in the reserve, will prepare the individual muster, retainer pay, and enrollment card, and will forward the same in quadruplicate to the Adjutant

and Inspector with the enrollment papers or reports of joining by transfer, together with available evidence tending to support claim of prior service.

(9) *The data concerning prior service, good-conduct medals, conduct markings, extraordinary heroism for the determination of the proper rate of pay* will be certified on the muster, retainer pay, and enrollment card by the Adjutant and Inspector, and the rate of pay, both at the termination of the prior marine service and for the current enrollment, will be similarly certified thereon by the Paymaster, Headquarters, Marine Corps.

(10) *Reports of active service, etc.*—The commanding officer will report in duplicate to the Adjutant and Inspector for transmission to the Paymaster, Headquarters, United States Marine Corps, in accordance with the instructions contained on the card, the performance of the required active service, confirmation in rank or rating, and charges against pay by reason of forfeiture, discharge, death, desertion, transfer, or any other cause.

(11) *Upon transfer* of enlisted men to Class 1 (*c*), 1 (*d*), 5 (*e*), or 5 (*f*), a notation will be made in their service-record books as follows: "Transferred (date) to (class 1 (*c*) or 1 (*d*)), Fleet Marine Corps Reserve," or "Transferred (date) to (class 5 (*e*) or 5 (*f*)), Marine Corps Reserve Flying Corps," as the case may be. Settlement of pay and clothing accounts will be made in the same manner as now obtains in the cases of men transferred to the retired list.

PAY AND ALLOWANCES.

1400

(1) *The annual retainer pay* is shown in the following table.

	Provisional.	Confirmed.		Remarks.
		Officers.	Men.	
Class I:				
(*a*)		2 months' base pay		
(*b*)			2 months' base pay	
(*c*)			$\frac{1}{3}$ base pay plus permanent additions.	Increased 10 per cent for men credited with extraordinary heroism in line of duty.
(*d*)			$\frac{1}{2}$ base pay plus permanent additions.	Increased 10 per cent for men credited with extraordinary heroism in line of duty or where mark in conduct for 20 years or more is not less than 95 per cent.
Class II	$12	2 months' base pay	2 months' base pay	
Class IV	$12	2 months' base pay	2 months' base pay	
Class V:				
(*a*)	$12			
(*b*)		2 months' base pay		
(*c*)	$12			
(*d*)			2 months' base pay	
(*e*)				Same as Class 1 (*c*).
(*f*)				Same as Class 1 (*d*).
Class VI	$12 in time of war.	2 months' base pay in time of war.	2 months' base pay in time of war.	

(2) *Additional pay for reenrollment.*—Reservists who enroll after the termination of an enlistment of four years' naval service, or who reenroll within four months of discharge from a four-year enrollment, receive an addition of 25% of their confirmed pay for each 4-year period of naval service, including service in the reserve, the total addition not to exceed 100%.

(3) *Pay on active duty.*—Reservists on active duty receive the full pay and allowances of their grades and ranks in time of war, and in time of peace, for training receive their retainer pay in addition.

(4) *Pay may be forfeited* by reason of failure to report for training or inspection or to report correct address.

(5) *Mileage,* transportation and subsistence to and from training will not be allowed when period of training is less than fifteen days.

(6) *Uniform gratuity.*—During each enrollment, upon first reporting for active service for training, in time of peace, reserve officers are credited with a uniform gratuity of $50 and reservists in enlisted grades with $30; upon first reporting for service in time of war or declared emergency, the credits are $150 and $60, respectively, less any other uniform gratuity credited during the enrollment. The gratuity in the case of an officer must be accompanied by a certificate from the commanding officer that the officer has provided himself with the necessary uniforms.

(7) *Transportation and subsistence.*—Reservists holding enlisted ranks ordered to active duty in time of war or national emergency, are furnished transportation and subsistence, in kind, from place of receipt of orders to place of duty, and upon return to inactive status are paid travel pay from place of duty to place of receipt of orders or bona fide home address, as they may elect, at the rate provided by law or regulations.

(8) Reservists holding enlisted ranks ordered to active duty for training are furnished transportation and subsistence, in kind, from and to the place of receipt of orders.

(9) Expense of transportation and subsistence to a recruiting station for the purpose of enrollment or reenrollment must be borne by the applicant.

INDEX.

A.

D.

O.

P.

T.

Time lost, entry of, in service record book, -- 935-936

○

Made in the USA
Columbia, SC
24 January 2020